I0031544

Dark Skies

Dark Skies addresses a significant gap in knowledge in relation to perspectives from the arts, humanities, and social sciences. In providing a new multi- and interdisciplinary field of inquiry, this book brings together engagements with dark skies from a variety of disciplinary backgrounds, empirical studies, and theoretical orientations.

Throughout history, the relationship with dark skies has generated a sense of wonder and awe and provided the basis for important cultural meanings and spiritual beliefs. However, the connection to darks skies is now under threat due to the widespread growth of light pollution and the harmful impacts that this has upon humans, non-humans, and the planet we share. This book, therefore, examines the rich potential of dark skies and their relationships with place, communities, and practices to provide new insights and understandings on their importance for our world in an era of climate emergency and environmental degradation.

This book is intended for a wide audience. It will be of interest to scholars, students, and professionals in geography, design, astronomy, anthropology, ecology, history, and public policy, as well as anyone who has an interest in how we can protect the night sky for the benefit of us all and the future generations to follow.

Nick Dunn is Professor of Urban Design and Executive Director of Imagination, the design and architecture research lab at Lancaster University, UK. He is Founding Director of the Dark Design Lab, exploring the impacts of nocturnal activity on humans and non-humans. Nick is a Director of DarkSky UK, promoting more sustainable relationships between the built environment and the night, as well as exploring ways to promote wider and inclusive participation with dark skies. He is the author of *Dark Matters: A Manifesto for the Nocturnal City* (2016) and co-editor of *Rethinking Darkness: Cultures, Histories, Practices* (2020). Nick is a keen nightwalker, has curated exhibitions, and has given invited talks at both literature and science festivals.

Tim Edensor is Professor of Social and Cultural Geography at the Institute of Place Management, Manchester Metropolitan University. He is the author of *Tourists at the Taj* (1998), *National Identity, Popular Culture and Everyday Life* (2002), *Industrial Ruins: Space, Aesthetics and Materiality* (2005), *From Light to Dark: Daylight, Illumination and Gloom* (2017), and *Stone: Stories of Urban Materiality* (2020). He is the editor of *Geographies of Rhythm* (2010) and co-editor of *The Routledge Handbook of Place* (2020), *Rethinking Darkness: Cultures, Histories, Practices* (2020), and *Weather: Spaces, Mobilities and Affects* (2020). His most recent book, about a Scottish medieval cross, is *Landscape, Materiality and Heritage: An Object Biography* (2022).

Dark Skies

Places, Practices, Communities

Edited by Nick Dunn
and Tim Edensor

Routledge
Taylor & Francis Group
LONDON AND NEW YORK

Designed cover image: Clatteringshaws, Galloway International Dark Sky Park, 16 May 2023. © Nick Dunn

First published 2024
by Routledge
4 Park Square, Milton Park, Abingdon, Oxon OX14 4RN

and by Routledge
605 Third Avenue, New York, NY 10158

Routledge is an imprint of the Taylor & Francis Group, an informa business

© 2024 selection and editorial matter, Nick Dunn and Tim Edensor; individual chapters, the contributors

The right of Nick Dunn and Tim Edensor to be identified as the authors of the editorial material, and of the authors for their individual chapters, has been asserted in accordance with sections 77 and 78 of the Copyright, Designs and Patents Act 1988.

The Open Access version of this book, available at www.taylorfrancis.com, has been made available under a Creative Commons Attribution-Non Commercial-No Derivatives (CC-BY-NC-ND) 4.0 license.

Trademark notice: Product or corporate names may be trademarks or registered trademarks, and are used only for identification and explanation without intent to infringe.

British Library Cataloguing-in-Publication Data
A catalogue record for this book is available from the British Library

Library of Congress Cataloging-in-Publication Data
Names: Dunn, Nick, 1974– editor. | Edensor, Tim, 1957– editor.
Title: Dark skies : places, practices, communities / edited by Nick Dunn and Tim Edensor.
Description: Abingdon, Oxon ; New York, NY : Routledge, 2024. | Includes bibliographical references and index.
Identifiers: LCCN 2023042783 (print) | LCCN 2023042784 (ebook) | ISBN 9781032528021 (hbk) | ISBN 9781032528038 (pbk) | ISBN 9781003408444 (ebk)
Subjects: LCSH: Light and darkness. | Light and darkness—Social aspects. | Light pollution. | Night—Psychological aspects. | Light—Psychological aspects. | Lighting—Social aspects.
Classification: LCC QC372 .D37 2024 (print) | LCC QC372 (ebook) | DDC 551.56/6—dc23/eng/20231011
LC record available at https://lccn.loc.gov/2023042783
LC ebook record available at https://lccn.loc.gov/2023042784

ISBN: 978-1-032-52802-1 (hbk)
ISBN: 978-1-032-52803-8 (pbk)
ISBN: 978-1-003-40844-4 (ebk)

DOI: 10.4324/9781003408444

Typeset in Times New Roman
by Apex CoVantage, LLC

Contents

Contributors

Kerem Asfuroglu is the founder of Dark Source, an award-winning lighting design studio driven by environmental values based in London and Dublin. Following his graduation from Wismar University—Architectural Lighting Design MA in 2010, Kerem has worked in the lighting industry for almost a decade before setting up Dark Source in 2019. He has been awarded the title of Dark Sky Defender by the IDA for advocating the importance of darkness through design. Some of his environmental lighting projects include the Plas y Brenin Outdoor Centre, Presteigne Dark Sky Masterplan, Newport Dark Sky Masterplan, Cloughjordan Ecovillage, Clwydian Range, and Dee Valley and Dark Sky Planning Guidelines for Cumbria.

Dwayne Avery is Associate Professor and Director of the communication studies program at Memorial University. His research explores the visual mediation of environmental pollution. He has published works on the branding strategies used by oil companies to mitigate small recurring oil spills. His current research explores contemporary media and the dark skies movement.

Louise Beer is an artist and curator, born in Aotearoa New Zealand. She now works between London, Margate, and Aotearoa. Louise uses installation, moving image, photography, and sound to explore humanity's evolving understanding of Earth's environments and the cosmos. Louise's experience of living under two types of night sky, the first in less light-polluted areas in Aotearoa and the second in more light-polluted cities and towns in England, has deeply informed her practice. Louise holds an MA art and science from Central Saint Martins and a BA (Hons) in fine art from Middlesex University London.

Therese Conway is a lecturer in the School of Geography, Archaeology, and Irish Studies at the University of Galway. Her research interests and publications relate to spatial planning, planning and development for tourism, niche tourism forms, and the role of social capital in rural development. Therese lectures widely in the discipline of geography and is also the director of the MA in rural futures planning and innovation program. Therese teaches specialist planning and tourism modules at BA and MA level and supervises MA and PhD projects. Dr Conway is a team member of the H2020 Ruralisation project and the Horizon Europe FLIARA Project.

Hannah Dalgleish is a researcher, communicator, and academic-policy engagement professional working in the science and society space. She completed a PhD in astrophysics (Liverpool John Moores University, 2020), followed by a postdoc at the University of Namibia working on astronomy for development. She then became an Oxford Policy Engagement Fellow exchanging knowledge with policymakers on light pollution in Armagh, Northern Ireland, followed by a short time as a Green Talents Fellow working on "rewilding the night" at the Sustainable AI Lab in Bonn, Germany. Hannah is now based at the University of Southampton working as a Knowledge Exchange Fellow.

Élisabeth de Bézenac is an architect, photographer, and PhD researcher in urban design at Lancaster University. Her research explores nocturnal environments in different parts of the world. She is currently developing a Night Design Guide, using fieldwork and collaborative projects to inform a framework to help urban designers account for their sites after dark in their creative process. She is a member of the Dark Design Lab, an interdisciplinary research team that addresses global urban night issues, exploring ways to mitigate artificial lighting and promote healthier rhythms of light and dark in urban areas.

Kimberly Dill is an environmental ethicist and Assistant Professor of Philosophy at Santa Clara University. Through her research and teaching, Dr Dill articulates a series of arguments in favour of biodiversity, wildness, forest, and dark sky conservation. She contributes to the literature in a distinctive way by critically evaluating empirical research conducted in environmental science and psychology, which demonstrates that there are a variety of worrisome psychophysiological harms associated with the loss of biodiversity, wildness, and natural darkness. This practical approach is situated within a broader ethical framework, which emphasises the importance of cultivating reciprocal, flourishing relationships with our more-than-human world. Furthermore, transformative, moral emotions—including (but not limited to) reverence, felt connectedness, awe, and wonder—she posits, motivate us to more fully participate in their successful conservation and regeneration.

Nick Dunn is Professor of Urban Design and Executive Director of Imagination, the design and architecture research lab at Lancaster University, UK. He is the Founding Director of the Dark Design Lab, exploring the impacts of nocturnal activity on humans and non-humans. Nick is a Director of DarkSky UK, promoting more sustainable relationships between the built environment and the night, as well as exploring ways to promote wider and inclusive participation with dark skies. He is the author of *Dark Matters: A Manifesto for the Nocturnal City* (2016) and co-editor of *Rethinking Darkness: Cultures, Histories, Practices* (2020). Nick is a keen nightwalker, has curated exhibitions, and has given invited talks at both literature and science festivals.

Tim Edensor is Professor of Social and Cultural Geography at the Institute of Place Management, Manchester Metropolitan University. He is the author of *Tourists at the Taj* (1998), *National Identity, Popular Culture and Everyday Life* (2002), *Industrial Ruins: Space, Aesthetics and Materiality* (2005), *From Light to Dark: Daylight, Illumination and Gloom* (2017), and *Stone: Stories of Urban Materiality* (2020). He is the editor of *Geographies of Rhythm* (2010) and co-editor of *The Routledge Handbook of Place* (2020), *Rethinking Darkness: Cultures, Histories, Practices* (2020) and *Weather: Spaces, Mobilities and Affects* (2020). His most recent book, about a Scottish medieval cross, is *Landscape, Materiality and Heritage: An Object Biography* (2022).

Rupert Griffiths is a cultural geographer with a background in architecture, urbanism, and microelectronic engineering and extensive experience as an artist, designer, and curator. His research considers the cultural imaginaries of urban nature, asking how urban and human-altered landscapes are legible as more-than-human ecologies. This contributes to a recalibration of urban imaginaries—and associated design practices—from human-centred to more-than-human. He uses practices of environmental observation, such as photography, phenology, and soundscape ecology, to develop imaginaries that capture and communicate the multiplicity of biological, non-biological, and technological rhythms through which our environments unfold.

Ysanne Holt is Emerita Professor in Art History at Northumbria University. Her research concerns the visual and material culture of the UK north, most especially of rural environments. Recent related publications include "Think Rural: Act Now: The State of the Countryside and Rural Arts Residencies in the 1970s and 80s", in Linda M. Ross et al.'s *New Lives New Landscapes Revisited*, 2023; "Sally Madge: acts of reclamation and renewal between site, studio, archive and gallery", with Matthew Hearn, *Journal of Visual Arts Practice*, Vol. 21, 2022 no. 3; and "Place on the Border and the LYC Museum and Art Gallery", in Tim Edensor et al.'s *The Routledge Handbook on Place*, 2020.

Ellen Jeffrey is an independent dance artist and researcher working with time-specific choreographic practices. She studied at Trinity Laban Conservatoire and University of the Arts Helsinki before completing her PhD at Lancaster University. Working collaboratively with local artists and communities to generate performances, films, workshops and writings, Ellen's work explores the capacity of dance and movement to attune to more-than-human timescales. She lives and works in Scotland.

Therésa Jones is Associate Professor in Evolutionary and Behavioural Ecology who explores the ecological impact of light pollution. Her research has demonstrated the largely negative effects of artificial light at night on the behaviour and physiology of multiple animal species and their communities. She works with external stakeholders to translate academic knowledge into real-world solutions that mitigate the impact of light pollution on wildlife. She has contributed to state and national light pollution guidelines in Australia and is a member of the Network for Ecological Research on Artificial Light and the Australasian Dark Sky Alliance.

Neha Khetrapal is Associate Professor at the Jindal Institute of Behavioural Sciences, O.P. Jindal Global University in India. She is interested in the interaction of religion with various aspects of environmental humanities. Within this domain, she is broadly interested in understanding whether traditions and religions afford imaginative possibilities for addressing the breakdown of human relations with the environment.

Yee-Man Lam has two identities: she is a research administrator as much as an independent researcher in Hong Kong. Soon after obtaining her doctorate degree from the Chinese University of Hong Kong, she worked as an Assistant Professor in the Department of English Language and Literature at Hong Kong Shue Yan University. Her publications can be found in the international peer-reviewed journals, such as *Ethics and the Environment*, *Visual Studies*, and *Nature+Culture*. Her research centres on environmental humanities, now with a specialised focus on the interplay between language and the environment.

Marty Lockett is a researcher in the effects of artificial light on ecological communities and a member of the Network for Ecological Research on Artificial Light. He works with local, state, and federal governments and industry to improve the quality and uptake of wildlife-sensitive lighting strategies and to bring a wildlife focus to outdoor lighting standards and regulation. Marty's broader ecological interests include the design, implementation, and analysis of ecological surveys; communicating ecological methods and data to a wide variety of audiences; and developing tools to assist environmental decision-makers at all levels.

Georgia MacMillan is an employment-based PhD scholar with National Parks and Wildlife Service, funded by the Irish Research Council. Her role as a Mayo Dark Sky Park development officer is an integral part of her research on the phenomenon of darkness and dark skies as a dimension of sustainable tourism in rural communities, using the case study site of Mayo Dark Sky Park. Georgia has a background in outdoor education and has been a dark sky community volunteer for nearly ten years.

Marie Mahon is a human geographer at University of Galway. Her main areas of research interest include the importance of the arts and culture to the rural; spatial justice, including rural spatial justice; governance of spatial planning, particularly issues of spatial justice and participatory democracy; and urban-rural relationships and development, including the nature of civic engagement and governance, place-based identity, and meaning and representations of rurality.

Helen McGhie is a photographer, a researcher, and Senior Lecturer in Photography at the Northern Centre of Photography, University of Sunderland, where she is also undertaking a practice-based PhD, funded by the National Productivity Investment Fund (AHRC). In partnership with Kielder Observatory (Northumberland, UK), her research contributes a model for how a creative practitioner and an astronomy organisation can undertake work with mutual benefits. She previously studied at the Royal College of Art (2012–2024) and has since exhibited nationally and internationally and contributed to publications, including *Madam and Eve: Women Portraying Women* (2018).

Natalie Marr is an artist and researcher based in Scotland, where she is completing a PhD at the School of Geographical and Earth Sciences, University of Glasgow. Her PhD research maps the cultural values of the Galloway Forest Dark Sky Park and its approach to dark skies practice. She makes work through photography, sound, writing, and collaboration with an interest in forms that centre experience and facilitate dialogue. She has presented and produced work nationally and internationally in contexts such as Sanctuary Arts Lab in Talnotry, Scotland, UK; the KTH Posthumanities Hub in Stockholm, SE; and the National Museum of Estonia, Tartu, EE.

Dan Oakley is the founder of Darkscape Consulting which is a dark sky and light pollution consultancy. He was the dark skies officer for the South Downs National Park Authority and was instrumental in securing a successful application for the South Downs to become an International Dark-Sky Association (IDA) Reserve in 2016. He established and coordinates the activities of the UK Dark Skies Partnership and is a founding director of DarkSky UK. He also sits as the current chairman of the Dark Sky Places Committee. Dan holds degrees in physics and wildlife management and a master's in environment policy and society.

Nona Schulte-Römer is a social scientist working at the Institute for European Ethnology at Humboldt-Universität zu Berlin. She has been researching light and darkness since 2008, when she began her sociological inquiry into the LED "revolution" in urban public lighting (*Innovating in Public*, 2015). She is lead author of the open access e-publication *Light Pollution—A Global Discussion* (2018) and part of the citizen science collective Nachtlichter documenting artificial light emissions in public spaces. Her current research examines public controversies about the environmental and health risks of light pollution, wireless telecommunications signals, and chemicals in the environment.

Taylor Stone is a Senior Researcher at the Institut für Wissenschaft und Ethik (IWE), University of Bonn. He received his PhD in ethics of technology from Delft University of Technology and has held postdoctoral and lecturer positions in industrial design, ethics of technology, and engineering ethics education. His research investigates the ethics of urban technologies, with a specialisation in lighting policy and design. His work in this area focuses on the ethics of light pollution and what it means to value—and ultimately design for—darkness.

Preface

The origins of this book arose through our shared love of dark sky places and the diversity of experiences and encounters they bring. Having individually written monographs on the subject of darkness and thoroughly enjoyed our collaboration on the edited collection, *Rethinking Darkness: Cultures, Histories, Practices*, we decided to convene three sessions, two related to Rethinking Darkness and one on Dark Skies: Borderlands, Cultures, Practices, at the Royal Geographical Society with Institute for British Geographers Annual International Conference 2021 in London. This enabled us to bring together a series of scholars and practitioners to share their engagements with dark skies from a variety of backgrounds and theoretical orientations. The sessions were a success, reminding us of how much more work there was to be done on the subject of darkness itself and inspiring us to bring this edited collection together. It includes contributions from some of those who presented at the conference. We discussed the scope of this book at length and given the vast array of disciplines that darkness connects with plus the manifold ways in which it is interpreted, experienced, contested, and examined, we decided to focus on the insights and methods that the arts, humanities, and social sciences can bring. In this regard, our book addresses a significant gap in current knowledge in relation to perspectives from these fields on the subject of dark skies and significantly extends our previous work on darkness. In building a multi- and interdisciplinary field of inquiry into dark skies, this edited collection aims to bring together engagements with dark skies from a variety of disciplinary backgrounds, empirical studies, and theoretical orientations.

Acknowledgements

We would like to express our considerable thanks to all the authors who contributed chapters to this edited collection. We are delighted with the breadth and depth of inquiries and creative engagements with dark skies that they represent. They have proved an attentive and supportive group who have been a pleasure to work with. Thank you, everyone. We would also like to thank Faye Leerink and the team at Routledge who championed this book from its conception and helped us bring it into being. We would particularly like to thank Research England and the E3 Beyond Imagination project which funded this publication. Nick would like to thank his family and friends for the love and support they gave through the encounters with dark places that made this book possible, especially Evelyn who is determined to be nocturnal because of all our conversations about non-human life after dark. He would also like to thank Tim for his generosity, energy, and enthusiasm in undertaking this project; he has been a wonderful collaborator once again. Tim would like to thank Nick for his kindness, astute efficiency, and convivial disposition. He would also like to thank many encouraging friends and especially his amazing family, Uma, Jay, and Kim, for their love and support.

In addition, Ysanne Holt would like to thank the artists Jo Hodges, Robbie Coleman, and Tim Shaw for the conversations about their engagements with dark skies. Kerem Asfuroglu, meanwhile, would like to give thanks to all the amazing people he has had the honour of working together in his adventures: Bob Mizon, Dani Robertson, Georgia MacMillan, David Shiel, Leigh Harling Bowen, Leigh Williams, Jack Ellerby, Lorayne Wall, and the local authorities and communities of Presteigne, Norton, Newport, and Cloughjordan.

Part 1

Introduction

1 Dark skies

Meanings, challenges, and relationships

Tim Edensor and Nick Dunn

Histories of dark skies

A fascination with the night sky is integral to the story of what it is to be human. The history of our relationship with dark skies is diverse and rich, a connection across space and time that has shaped and been shaped by society, culture, and religion, as well as science. Beyond the astronomical, scientific understandings about the universe, the stars, planets, and moon have proved inspiration for artists, poets, and philosophers. As a realm in which we search for meaning, dark skies have been integral to how we commune with our hopes and fears. And while there is no definitive theory as to how or why humans started to relate to dark skies, a growing body of evidence across a range of disciplines suggests that it is plausible that it began at about the same time as recognisably modern humans evolved, around 70,000 years ago (Clark, 2020, p. 4). Before the advent of modern science, the relationship between an individual and the night sky was typically immediate and powerful. With no light pollution to obstruct the view, the stars were conspicuous, and cosmic understandings were shaped by common beliefs that the conditions of dark skies revealed what was happening on Earth.

In their significant work on exploring Australian Aboriginal meanings and uses of dark skies, Karlie Noon and Krystal De Napoli (2022) investigate cultural interpretations and practices that go back at least 65,000 years. To Western thinking, stories based on astronomical figures may appear mythical; however, they contain much that is practical. Noon and de Napoli demonstrate the highly sophisticated, complex, and diverse ways in which different Indigenous communities have used the star-studded sky to provide a guide for navigation, mark times for hunting, journeying, and collecting, and demarcate seasonal ceremonies and rituals. Across millennia, elements in the night sky have been scripted into the land, generating tales that explain the mythic origins of landforms and trigger the enaction of travel, song cycles, dance, and art. Importantly, these oral traditions avoid the reifications generated by writing down authoritative knowledge, with understandings and practices remaining dynamic, evolving, protean, and multiple.

More specifically, Noon and de Napoli (2022, p. 43) show how knowledge from the skies has informed "interrelations between ecological, medicinal, celestial and technological knowledge", understandings that are simultaneously relational,

DOI: 10.4324/9781003408444-2
This chapter has been made available under a CC-BY-NC-ND 4.0 license

practical and cyclic, that mark weather patterns (through star scintillation, for instance), the fruiting of edible plants and the related migratory routes that lead animals to consume food sources.

Such networks of associations spread across space and time but are necessarily situated amongst particular Indigenous groups and spaces. And while the ever-changing positions in the firmament mark out daily, seasonal, and annual fluctuations, they also inscribe much longer, millennium-spanning changes. For shifting alignments in the sky, over 13,000-to-26,000-year cycles are registered through oral traditions that testify to the longevity of knowledge and the ancient witnessing of celestial processes. These extensive temporalities are also underlined by the continuing importance of faint dark sky constellations; those gaps in which stars cannot be seen are sites of enduring knowledge. With the advent of light pollution, many such cosmic spaces can now rarely be discerned. Best known of these dark realms is the widespread phenomenon known as the Dark Emu, a bird-shaped hole in the Milky Way and the focus of diverse myths from different Aboriginal nations. By following the progress of the bird across the sky throughout the year, the most auspicious time to collect emu eggs, visit waterholes, and move camps are identified.

Noon and de Napoli consider that such ancient, intimate knowledge of the cosmos can inform contemporary approaches to environmental change. These land-sky associations strongly resonate with the emergence of contemporary relational and vitalist thinking across the social sciences. Moreover, they regard preservation of this knowledge as essential, for it constitutes an element of "space heritage" that encompasses a vast cosmological archive of knowledge, stories, and practices. At present, the more intimate areas of knowledge are not disclosed to those outside Indigenous communities but retained by the appropriate knowledge keepers.

In later prehistoric times, a plethora of Neolithic stone circles, burial chambers, and cairns have been construed by many archaeologists as relating to astronomical events such as solstices, lunar and solar phases, and the location of star constellations. Indeed, Ronald Hutton (2018) details the rise of archaeoastronomy as a means to explore whether such sites, most notably Stonehenge, were devised as ancient observatories and their design informed by advanced celestial knowledge. Such studies remain inconclusive. However, the significance of these cosmic associations is central to a range of post-Christian countercultural, pagan, and new age groups as sites of ancient wisdom that stand in opposition to conventional academic knowledge and scientistic ways of thinking.

In considering the numerous myths and legends that focused upon the night sky in ancient times, Patrick McCafferty (2018, p. 24) conceives the sky to be "a tapestry embroidered by human imagination", containing stars that "might be viewed as bright dots that can be joined to create a skyscape of characters and imaginary objects". While these patterns have diminished as objects of scientific classification, their mythic potential lingers in popular astrological imagery wherein celestial bodies and their movements influence the individual and mark their place in the universe and Earth through the star signs system. Earlier conceptions identify stars, moon, sun, and planets as potent cosmic forces that delineate mythic deities and events, some of which might influence earthly occurrences. For instance, the moon for Romans was the two-horned goddess Luna, for the Maya it was a goddess or rabbit, while for Hindus its eclipse was caused by it being periodically swallowing by the demigod Rahu. Myths have also congregated around comets, meteors, and fireballs in ancient Greek and Roman cultures, the Bible, the Ramayana, the Norse Ragnarok, and the Persian Shahnameh, typically serving as signs of superhuman and supernatural battles, travels, and chases or as portents foretelling of the advent of war, famine, and plague. In focusing specifically on darkness, ancient Greeks practised necromancy and Dionysian and sacrificial rituals in the dark that were associated with the wholly lightless conditions of the Underworld, in contradistinction to the variable darkness of earthly night-time (Boutsikas, 2017).

April Nowell and Nancy Gonlin (2020) investigate how archaeological research can reveal how everyday, sacred, and productive nocturnal ancient cultures that fashioned artefacts, symbols at important sites were carried out, refuting the overemphasis on diurnal practices and beliefs. In advancing understandings about the integral place of dark skies in prehistoric and later cultures, their edited collection (Gonlin and Nowell, 2018) explores the "archaeology of the night" through essays that focus on the diverse, complex nocturnal historical practices and understandings of Polynesian, Native American, Indian, South American, African, and Arabian cultures, collectively refuting any overdetermined, universalist conceptions about the meanings of dark skies and darkness. Indeed, a host of recent accounts focus on the very particular cultural historical practices and meanings organised around dark skies.

Guy Bordin (2020) shows how before their conversion to Christianity and its binary notions of light as good and darkness as evil, North Baffin Inuit of the eastern Canadian Artic construed darkness as conducive to sociability, storytelling, play, rituals, and festivity. In Maori cultures, such cosmological traditions continue, with rituals of praying, drumming, and chanting into the night (Lees, 2022). In her chapter for this book, Neha Khetrapal provides a fascinating anthropological account of the dense Indian pre- and post-Vedic myths that identify the patterns formed by constellations, stars, and celestial groups, storied imaginaries that have influenced the production of material cultures. These astronomical interpretations were changed to accommodate agricultural cycles as they shifted from hunting and gathering and were aligned with the phases of the moon and menstruation, embodied in pottery, seals, and figurines that guided these and other economic and cultural practices.

Strikingly, Khetrapal shows how the Pleiades, embedded in ancient Indian religious and mythic systems, represents the story of the Seven Sisters, a commonality that is separately configured across cultures from widely diverse spatial and historical contexts. Significantly, she also demonstrates that the previously dominant lunar myths that centre upon female celestial characters and female power, often embodied in the figure of *Śakti*, were relegated in importance with the arrival of Aryan myths that privileged sun gods. The solar replaced the lunar, and the daytime triumphed over the night in the gendered reconfiguration of belief and myth.

The modern urge to look at our prehistorical relationship with dark skies has its origins in the work of Alexander Marshack, an American journalist turned archaeologist. Marshack's interest was sparked on 4 October 1957 when Sputnik 1, the world's first spacecraft, was launched by the Soviet Union. Yet what distinguished Marshack from many of his contemporaries was not just a keen interest in the technological advancements that had made this possible but something more primal— he was fascinated by what had driven humans to desire to "touch" the night sky. In his attempt to explain the emotional pull of dark skies, Marshack (1972) sought to identify when our relationship with them began. This research took him back tens of thousands of years ago to the time prior to civilisation and agriculture, to when humans lived in hunter-gatherer communities.

Other forces have sought to diminish the potent myths and practicalities that have surrounded dark skies. Indeed, as long ago as 2,400 years ago, in Greek philosopher Plato's Book VII of *The Republic*, he hypothesised that our eyes were formed for the study of the night sky. Rather than being awestruck by its wonder, he encouraged us to understand the order within its celestial arrangements. Much later, during the 16th and 17th centuries, astronomers profoundly altered ancient understandings, their scientific discoveries creating a state of uncertainty that was characterised by the German sociologist Max Weber as "disenchantment". The emergence of this astronomical knowledge resonated with his view that the Enlightenment initiated the process through which our enchantment with the world was replaced by scientific rationality. Weber argued that we lost something of the sublime that inspired our imaginations and emotionally connected us to the night sky and other wonders of nature (Jenkins, 2000, pp. 11–32), a loss that affected us on a fundamental level. A critical change concerned our perception of the night sky, which could no longer be viewed as a firmament, effectively a two-dimensional pattern of stars and planets, but rather a three-dimensional realm of inconceivable size. Yet this disenchanting disconnection brought new, different kinds of awareness, as scientific discoveries made the cosmos—previously regarded as unknowable— increasingly tangible. This idea was both exhilarating and alarming, and the combination of pleasure and fear was irresistible to philosophers who had long been preoccupied with what they referred to as the sublime, a relationship which continued to endure. The rebirth of astrology in the early 20th century, for example, fascinated the psychoanalyst Carl Jung (1976) who considered the night sky to be the ideal psychological mirror of our innermost thoughts.

Throughout history, the stars have provided humans with increased understanding about the world and the universe within which it sits. This has enabled significant scientific discoveries to be made and contributed toward accurate systems of navigation. Yet the dark skies movement, besides being informed by scientific astronomical explorations and spectacles of the night sky is part of a process of re-enchantment with dark, nocturnal space, as amply demonstrated by the chapters in this book. This is further borne through the recognition that ancient myths can tell us about relating to the dark skies in more sustainable, less rational and enchanted ways, allowing us to rediscover more varied, complex perspectives through which to experience them.

The multiple meanings of darkness: its ambiguities and its relationship to light

While mythic stories about the cosmic significance of the night sky have forever circulated, happenings on the nocturnal landscapes of Earth have been shrouded in different kinds of beliefs and legends. Before the advent of widespread artificial illumination, darkness shaped the enduring human experience of the night, largely ameliorated only by minimal forms of illumination, such as fire, sputtering candles, and rushlights. It is little wonder that in certain cultural, historical contexts, the dark night became associated with negative attributes, saturated with fear and superstition. As Galinier *et al.* (2010, p. 820) contend, since medieval times, darkness has symbolised "pagan obscurantism—deviancy, monstrosity, diabolism". Sinister hobgoblins, ghouls, ghosts, witches, demons, dark elves, and the devil himself might be discerned in murky, shadowy forms. Superstitious beliefs were supplemented by influential biblical passages that underlined binary distinctions between a malign darkness and a godly realm of light. Yet perceptions of the dangers posed by darkness were not merely imaginary. As Roger Ekirch (2005, p. 123) details, numerous very real hazards lay in wait for those who ventured out after nightfall, who might have to negotiate piles of rubbish, ditches full of waste, and overhanging timbers, not to mention violent criminals. And those venturing beyond the city might stumble into "fallen trees, thick underbrush, steep hillsides and open trenches".

Schlör (1998, p. 57), meanwhile, emphasises the dominance of light over dark in considering the urban night, "our image of the night in the big cities is oddly enough determined by what the historians of lighting say about *light*. Only with artificial light, they tell us, do the contours of the nocturnal city emerge: the city is characterized by light". It is perhaps unsurprising with this backdrop of understanding that the role of darkness in urban areas is typically met with negative perceptions. Extensive illumination has been championed because it affords visibility and orientation, deterring potential wrongdoers, and so darkness remains associated with the fearful and dangerous (Brands, Schwanen and Van Aalst, 2015). Yet this is highly ambiguous; there is growing awareness that light renders potential targets visible and disarms visibility outside areas of bright illumination. Despite powerful orthodoxies that greater levels of street lighting reduce crime, no evidence shows this to be the case. Indeed, as Green *et al.* (2015) remark, lower levels of light make the identification of likely victims more difficult to discern, and they further contend that the impact of street dimming schemes promote an increase neither incidences of crime nor road traffic accidents.

Dystopian nocturnal imaginaries have informed apocalyptic visions of the future. In exemplifying an early representation of end times, Tiffany Francis cites Byron's extraordinarily poem from 1816, *Darkness*, in which he portrays a cataclysmic ending as the light of Earth is extinguished and life sputters out, returning to the darkness of the universe. Such scenarios are reiterated in a range of contemporary fictional and filmic science fiction works, as darkness descends across the Earth, concealing alien and unseen threats, sparking resonances with ghost stories and horror movies.

Despite these negative representations of darkness, Robert Williams (2008, p. 514) adopts a broader perspective, contending that dark spaces are "constituted by social struggles about what should and should not happen in certain places during the dark of the night". For though orthodox conventions about the desirability of banishing darkness are often articulated and enacted by the powerful, who regard gloomy spaces with anxiety and suspicion, others have sought darkness positively, regarding it as conducive to a range of cultural practices. In the dark, persecuted minorities, marginal groups, and the lower classes may escape domineering masters and carve out time and space so that they may temporarily achieve "freedom from both labour and social scrutiny" (Ekirch, 2005, p. 227). Galinier *et al.* discuss how the Mesoamericans and Andeans who escaped the violence of Spanish imperial power confined their "indigenous knowledge and practices to the hidden recesses of the night" (2010, p. 828), and Palmer (2000) mentions how African-American slaves forged a collectivity in the darkness. As illumination has expanded it has been resisted by "the traditional inhabitants of the night: servants, apprentices and students . . . tavern visitors, prostitutes", musicians, mystics, and bohemian pleasure-seekers (Koslofsky, 2011, p. 278). Criminals and political and religious dissidents also seek gloom, for "under cover of darkness spirits roam, plots are hatched, raids are executed and the dangerous escape" (Lees, 2022, p. 153).

These diverse and often divergent understandings underline the contested, ambiguous, and multiple meanings of darkness and the different subjective feelings that it raises. Those who seek darkness to engage in quiet, reflective meditation contrast with others who feel fear and suspicion about what may lurk unseen; a brightly illuminated urban commercial area may lure shoppers but might be regarded as a realm of surveillance and exclusion by more marginal residents of the city. Such different perspectives call for a thorough reassessment of the multiple, changing, and contested qualities attributed to darkness, and the chapters in this book provide an array of suggestions about how we might more productively encounter dark spaces.

The relationship of darkness with light, the advent of artificial illumination

It is difficult to imagine the affects and meanings that circulated amongst human communities in the absence of artificial light. The use of fire would have created pools of light, offering some safety against the threat of predatory creatures and reducing fear of the unseen. Firelight would have stood out against the blackness of the darkest nights and would have profoundly intensified the experience of numerous festivals. The Celtic festivals of Imbolc, Beltane, Lughnasadh, and Samhain, which marked the annual equinoxes and solstices, were celebrated with home fires, bonfires, and flaming torches, elements subsequently appropriated and celebrated by contemporary Wiccans and neo-pagans. Later, the darkness was periodically punctuated by fireworks and the lighting of beacons, and the widespread illumination provided over several nights by the lights of the Hindu festival of Diwali. Besides these modest practices to animate the night, the dark landscape itself could contain lights of uncertain provenance, notably marsh gas or will-o'-the-wisp that summoned up supernatural forces, as well as celestial phenomena such as meteor showers and comets. The mystery of such illuminated forms in the night sky do not merely belong to the past; they are contemporaneously expressed by inexplicable celestial phenomena often conjured as unidentified flying objects, signs of the spacecraft of technologically advanced visitors from other galaxies.

Undoubtedly, as these examples demonstrate, and as Wolfgang Schivelbusch (1988) claims, perceptions of artificial lighting at night have consistently fused the literal and symbolic. Yet the reduction in darkness perpetrated by illumination has diminished its power to contain the mythical, the inexplicable and the sacred. This underlines the potency of gaslight and subsequently electric light in coming to define "a new landscape of modernity" (Nasaw, 1999, p. 8) that transformed the city from a dark realm into a brightly illuminated, more tightly regulated space. Craig Koslofsky (2011) calls the expansion of social and economic activity into the night through the spread of illumination "nocturnalisation", while Murray Melbin (1978) refers to the "colonisation" of the night. As Bille and Sørensen (2007, pp. 272–273) note, these processes are further underpinned by the metaphorical quest by Enlightenment scientists and thinkers to "shed light on all things" in the pursuit of "truth, purity, revelation and knowledge"; the ideals of "illumination, objectivity and wisdom" are synonymous. Such ideals bolstered colonial Eurocentric ideas that Africa was the "Dark Continent", replete with animist and idolatrous faiths, ignorance, and barbarism, "primitive" qualities that could be overcome by the colonial civilising mission, malign associations that were transmuted to the slums of "darkest London".

Besides serving to demonise darkness as its opposite, illumination also contributed to the production of a phantasmagoric realm (Collins and Jervis, 2008) that defamiliarised, excited, and transcended ordinary ways of sensing. A host of unlit structures and spaces were imperceptible while illuminated buildings stood out, new colours were introduced to the nightscape, light was refracted in puddles and windows, and scale and proportion were difficult to determine. Much modern space was thus transformed "into a perceptual laboratory" (McQuire, 2008, p. 114), becoming a dream-like, often fantastical realm, especially with the advent of illuminated shop windows, pleasure gardens and amusement parks, and seasonal light festivals. In one sense, this extensive illumination produces a new kind of nocturnal sublime that persists, as Jean Baudrillard (1986, p. 51) observes,

> There is nothing to match flying over Los Angeles by night. A sort of luminous, geometric, incandescent immensity, stretching as far as the eye can see, bursting out from the cracks in the clouds . . . This [city] condenses by night the entire future geometry of the networks of human relations, gleaming in their abstraction, luminous in their extension, astral in their reproduction to infinity.

In contemporary times, light festivals continue to transform and enchant nocturnal landscapes in which unheralded or unnoticed features are brought to attention, peculiar illuminated sculptures appear, odd atmospheres are generated by the judicious use of coloured lights and areas of darkness, and buildings are dematerialised by shape-shifting projections (Edensor, 2017). Such creative techniques defamiliarise places while drawing attention to their specific qualities, while promoting interactivity and conviviality. Yet such installations possess most power when positioned against a dark background, made more potent in the absence of other ambient illumination. Thus, darkness remains central to such experiences.

One further point needs to be made about the reconfiguration of the relationship between light and dark. For in addition to its aesthetic pleasures, and besides its economic, way-finding, and regulatory advantages, hierarchical notions were consolidated wherein brightly lit commercial and wealthier residential areas came to contrast with the dark neighbourhoods of the poor. As Joachim Schlör (1998, p. 65) notes, the brighter the light in the centres, "the more starkly do the outlines of the darker regions stand out". Darkness became "a symbol and a determinant of urban differentiation" (Otter, 2008, p. 335) in which poorer, gloomy realms were "disregarded and disparaged in relation to the new" (Brox, 2010, p. 104), seemed to comprise "another country" to the brightly illuminated shop windows, signs, theatre entrances, homes, and pubs in the commercial centres. Such distinctions have also been etched through the racialisation of space, as Harrison exemplifies in his account of the city of Rocky Mount in North Carolina, where the distribution of electric lighting was iniquitously deployed across urban space. Deemed necessary to prevent crime in commercial areas and white districts, illumination was regarded as unnecessary for black areas.

As Savela (2023) notes in his discussion of the Finnish city of Turku's redevelopment schemes, the deployment of light to mark out areas of privilege, exclusivity, and power endure, with brilliantly illuminated advertisements predominating in a way that would not be possible in the more cluttered visual array of the daytime. And in East London, the low-grade illumination of industrial areas, retail centres, and working-class housing estates continue to dramatically contrast with the brilliant corporate logos shining atop the large towers of Canary Wharf and the subtle, minimalist lighting schemes of recent upmarket residential developments (Ebbensgaard and Edensor, 2021). Strikingly, hierarchical differentiations are being inverted; in an over-illuminated world darkness may now be sought as a marker of status that differentiates wealthy from poorer, light-bound subjects and spaces (Sloane, 2016).

As the chapters in this book reveal, despite its modern construal as a realm of lurking danger, criminality, ignorance, and backwardness, darkness is now being more substantively re-enchanted. Perhaps as they have become less central to human experience, the loss of the night has prompted a reconsideration of the virtues of darkness, and this, in turn, has encouraged the ongoing, proliferating designation of dark sky places as realms in which multiple experiences of gloom may emerge. For many, the night is too bright, too indistinguishable from diurnal apprehensions, because it has been extensively over-illuminated, as we now discuss.

The loss of the night: darkness and over-illumination

The binary ways in which the relative values of dark and light are mapped onto geographical understandings of space after nightfall remains a major barrier to advancing the re-evaluation of dark skies. The persistent modern urge to flood nocturnal space with light inheres in the expansion of illumination, which has transformed the atmospheres of places at night, giving rise to new forms of labour and enabling social activities to wrap around the clock. Consequently, a prevailing tendency regards urban electric lighting as integral to their vibrancy and character, while in contrast, normative perceptions of rural places is that they will be satisfyingly dark the further they are away from urban areas. These dual notions of pristine "natural" darkness in rural areas and the bright lights of urban spaces are powerful in shaping how we apprehend whether places at night should be illuminated or not. Yet this thinking is reductive. While protecting dark reserves is important, tackling light pollution concerns all forms of landscape such is its pervasive growth. These difficulties in understanding are exacerbated by evaluating the value of dark skies (Henderson, 2010; Gallaway, 2015) beyond these long-held beliefs about light and dark space (Gallan and Gibson, 2011). Yet the articulation of such notions is geographically variable.

In her chapter for this book, Yee-Man Lam explores the diverse meanings of dark skies across contemporary cultural and geographical contexts. Drawing upon the results from putting the phrase "dark sky" into an internet search engine, she finds that since 2008, the most common links are to popular cultural forms, such as songs, novels, films, and commodities, and to dark sky settings and events. By contrast, an internet search across the same period for solely Chinese references to "dark sky" disclosed no findings at all, although there were plenty of links to "starry sky" and "night sky", the later exclusively focused on pop songs that include the phrase in their titles or lyrics, while "starry skies" brings up a broader range of popular cultural productions. Lam draws out two conclusions from her research. First, that the links to "dark sky" in non-Chinese contexts are deceptive in that what is really sought are the stars that illuminate the dark sky, not darkness itself. This may represent the dominance of astronomers as initiating advocacy for the assignation of dark sky places, but as the chapters in this book demonstrate, dark skies and darkness are being increasingly valued for the many other potentialities they promise. Second, in exploring the Chinese dearth of links, the search is narrowed to Hong Kong, where intense nocturnal illumination eliminates any possibility of dark skies, and culturally speaking, light remains integral to Hong Kong's entrepreneurial identity, with darkness associated with crime and poverty. The discussion critically reveals that the positive values increasingly associated with dark skies have not gained saliency in many parts of the world. Lam's example thus further demonstrates how negative associations about darkness persist, as Kumar and Shaw (2019) show in disclosing how illumination remains a signifier of modernity in rural India.

The advantages of illumination have been crucial to the development of towns and cities after dark, and we should not seek to refute how light brings order and regulation, safety, ease of movement, accessibility, and way marking to formerly dark spaces. Yet its unfettered, global growth across maritime and land-based realms has heralded the advent of a serious, malign ecological, health, and aesthetic effects. In seeking to extend awareness of the spread of light pollution, in 2001, a team of astronomers have created the first atlas of light pollution (Cinzano, Falchi and Elvidge, 2001). More recently, new technologies that have produced depictions of the Earth at night, such as NASA's composite satellite imagery, although according to Pritchard (2017) this has reinforced binary ideas of light and dark, including normative ideas about where both conditions are located.

Light pollution is, however, increasingly being recognised as a global challenge (Davies and Smyth, 2018), the cascading effects of which remain unknown. Historically, problematic concerns about of light pollution have been connected to urban areas (Meier *et al.*, 2014) that contained the greatest concentrations of outdoor artificial illumination. However, artificial light's impact now extends far out into rural (Falchi *et al.*, 2016) and maritime areas (Smyth *et al.*, 2021) that might be far from urban places. This has already led to an "extinction of experience" (Pyle, 1978; Soga and Gaston, 2016) in many places, in which people are unable to access and thereby appreciate dark skies. This loss has major implications, not least because it further complicates the challenge to retain awareness of the value of dark skies and our relationships with them, alongside the benefits for other species and the planet.

The impacts of light pollution have been further problematised by the recent rollout of LED lighting across the globe that has introduced initially unforeseen and unintended consequences. For example, although the economic and technological efficiency of LED lighting technologies have driven their implementation, the cost benefits have masked how energy usage for outdoor lighting and artificial night-time brightness both continue to increase (Kyba, Hänel and Hölker, 2014). In addition, this contemporary transition is generally shifting the colour of night as tonalities of the light source toward increasingly blue-white and directional light than previously experienced which has greater impacts on ecosystems, as Jones and Lockett explore in their chapter for this book (also see Pawson and Bader, 2014).

Yet as Taylor Stone emphasises in his chapter, a quest to restore historical experiences of dark skies through a purist approach is unrealistic. What is required is the rebalancing of the relationship between light and dark in diverse situational contexts; what Stone refers to as repairing the relationship between urban inhabitants and darkness by incorporating lighting designs that are attuned to a wildness that does not exclude humans. While some campaigners call for standardised measurements and placements in efforts to regulate the damage caused by light, others argue for more cautious policies. A modest approach is recommended by Jones and Lockett, who identify the complexities of each situational context and the diverse theories that contest levels of harm perpetrated by excessive illumination on humans and non-humans. In their chapter in this book, Tim Edensor and Dan Oakley present further ways for challenging normative views concerning light and dark through their examination of festive illuminations in dark sky places. They demonstrate how creative, sustainable, and place-specific interventions can unlock more progressive approaches to engaging with gloom and place. Edensor and Oakley identify the need for more influential guidance and explore how inventive, yet less intrusive illumination might enhance the qualities of places after dark in urban and rural settings.

However, as more evidence is revealed about its harmful effects, the presence of light pollution is not simply a matter for cultural perceptions and social behaviours. There is a growing body of evidence concerning the problems light pollution is producing for human health (Chepesiuk, 2009; Falchi *et al.*, 2011). Exposure to LED lighting, increasingly common in both indoor and outdoor environments, has been linked to chronic sleep and circadian disruptions. Altering natural cycles of light and dark directly affects the rhythms of our bodies and minds with serious health consequences including cancer, cardiovascular disease, diabetes, and obesity (Rijo-Ferreira and Takahashi, 2019). Artificial light can make our species less healthy and, in some cases, cause very serious reactions (Levin, 2019). As we now discuss, this damage extends to destructive impacts upon non-humans and the ecological spaces they occupy.

Effects of over-illumination on non-humans

Before the advent of humans on Earth, dark skies were used for navigation by other species, culminating in the evolution of distinct diurnal, nocturnal, crepuscular, and cathemeral creatures and the biogeographies in which they live. Dark skies remain crucial for navigation by other species, including night-migrating birds, seals, dung beetles and certain species of moths (Foster *et al.*, 2018). Yet this wayfinding is only one aspect of how other animals relate to dark skies. Most mammal species are nocturnal, but many are diurnal, active during the day; crepuscular, mostly active around twilight; or cathemeral, active during hours of both daylight and darkness. These various strategies for regulating activity over a 24-hour cycle are related to evolutionary adaptations to light or semidarkness.

The nocturnal activity of non-humans seethes through the night, with certain realms becoming extraordinarily animated after darkness falls. Dixe Wills (2015) describes the swirling din and movement he experiences during a nocturnal visit to the Isle of Skomer in Wales. Here, having spent hours hunting at sea, hundreds of thousands of Manx shearwaters return to their burrows to feed their chicks, choosing the darkest conditions in which to land so as to thwart the threats posed by the squawking, predatory seagulls that await their arrival. As Lees (2022, p. 14) relates, in the sea, "the shift to twilight triggers squid to rise, jellyfish to descend, pink mamao to become paler and more mottled as they sink, and luminescent plankton to flash". Botanically, too, photoreceptors in leaves and petals flood flowers with perfumes to attract pollinators while territorial animals lay down scents. These examples provide a small indication of the sheer complexity of non-human responses to light and dark, integral to genomes that have evolved over millions of years. Such flourishing in under great threat from over-illumination.

Living in cities and towns can have many benefits but there are also negative consequences. Urban pollution, for example, manifests in distinct forms and across

different mediums, notably through air, noise, and waste. These represent significant challenges for urban life and can have serious impacts on human health and
the wider environment. However, until recently, the problems caused by light pollution have garnered little relatively attention and are only starting to enter the public imagination as a serious problem (Drake, 2019). The introduction of artificial
illumination and its subsequent development across urban spaces and its extension
and colonisation of rural landscapes has become so ubiquitous that many rarely
think to question its presence and deployment in the environments we inhabit.

Its effects are not merely limited to humans. Darkness is integral to biodiversity and its disappearance is having far-reaching consequences for other species
(Sánchez-Bayo and Wyckhuys, 2019; Gaston, Visser and Hölker, 2015). This
means that places where dark skies are protected are becoming more and more
important to planetary health. Nocturnal rhythms and behaviours of flora and fauna
are disturbed by artificial light as the sensory capacities of nonhuman creatures
to breed, feed, prey, and migrate are impacted (Haim, Scantlebury and Zubidat,
2019). There have been high-profile cases of the many migrating birds who, disorientated by electric lights, become victims of "fatal light attraction" and crash
into buildings and other engineered structures (Van Doren *et al.*, 2017). Furthermore, the encroachment of artificial light at night into dark places also disrupts the
rhythms of insects (such as moths, fireflies, and beetles), bats, salamanders, and
migrating sea turtles (Rich and Longcore, 2006). Coastal and maritime lighting is
also responsible for suppressing the colonisation of certain species yet promoting
harmful others that foul harbours and coastlines to flourish, precipitating ecological
collapse (Davies *et al.*, 2015) and thus promoting demands for the designation of
marine dark sky parks.

Clearly, light pollution can create critical, if not downright disastrous effects
in disturbing the lives and ecosystems of the planet. In their chapter for this book,
Therésa Jones and Marty Lockett provide a more substantive account of these
extremely diverse effects on a host of non-humans, before considering how we
might mitigate these problems through a better balance between human desires
for nocturnal brightness and its impacts on other species. Kimberly Dill's chapter focuses on how woodland environments, crucial for sustaining biodiversity,
carbon sequestration, and oxygen production, are being enormously harmed by
excessive illumination, which affects their non-human inhabitants in numerous
ways. She provides a compelling argument for the restoration of natural darkness
to biodiverse, forested ecosystems, citing the impact that light pollution has upon
such places, compromising their ability to sequester carbon and thereby indirectly
contributing to global climate change. In addition, Dill reminds us that immersion in dark, biodiverse forests has been, and remains pertinent to, transformative
experiences for humans which the loss of darkness profoundly diminishes. In their
chapter in this book, Rupert Griffiths, Nick Dunn, and Élisabeth de Bézenac further
bridge between the human and non-human experiences of dark places. They discuss how use of unattended sensor methods, photography, and walking can be used
to create thick descriptions of places after dark. By apprehending darkness in different ways, the descriptions move between systematic environmental observation,

imaginative interpretation, and environmental and bodily rhythms and sensation. Through this, Griffiths, Dunn, and de Bézenac give expression to different nuances and values of darkness, then reflect on how such knowledge of the lived experience of humans and non-humans might inform design strategies that construe the urban environment as a more-than-human ecology.

To reiterate our earlier discussion, a reduction in the dynamic non-human agencies that swarm under dark skies are contributing to the human experiential extinction (Soga and Gaston, 2016) of the night. We illustrate this by referring to Chris Yates's (2012) account of nocturnal walking through the rural south England, in which he charts the gradual loss of the sights, sounds, and movements of many of the non-humans he formerly encountered but are now seldom witnessed. His elegiac account of the depletion of moths, birds (owls, skylarks, nightjars, nightingales), mammals, and insects from the nocturnal countryside that he regularly encountered a few short decades ago draws attention to the vicious impacts of the chemicals routinely sprayed on crops and fields, the removal of hedgerows, and the effects of over-illumination that produce this quieter, less lively night-time landscape.

Before this loss, such experiences were able to promote a renewed appreciation for the sheer otherness of the non-humans with whom we share the planet. In contextualising our specifically human capacities for sensing and moving under dark skies—discussed later—it is salient to consider the myriad, often unimaginable and uncanny ways in which non-humans deploy their sensory aptitudes after nightfall. Bats locate prey through echolocation, snakes through their sensitive tongues, sharks swim through dark seas via electrolocation, badgers and dogs are primarily guided through dark space by their extraordinary ability to apprehend the world through smell, and moles sense their utterly dark environs by detecting vibrations.

The rise of the dark sky movement

It is through an urgent desire to preserve a relationship with dark space, with the starry skies it promises, the creatures who emerge, the aesthetic experiences it encourages and the communal encounters it fosters that the dark sky movement has gained in popularity. Such positive experiences, along with growing awareness of the damage caused by over-illumination are contributing to the expansion of designated dark skies and spaces across the world.

In summarising the virtues of the dark night sky, Terrel Gallaway (2015, p. 280) identifies how it has been a "source of aesthetic, scientific and spiritual inspiration . . . a natural resource, a scenic asset and part of humanity's cultural heritage", as well as signifying more ecologically sustainable environments (Sutton and Elvidge, 2015). Such values inform the rise of the dark sky movement, a worldwide campaign whose overarching goal is to decrease light pollution to preserve the natural night sky. Although the primary aim was to enable access to starry skies, subsequent research has proven that there are other significant advantages to reducing light pollution in the environment, improving the health, wellbeing, and safety of humans and non-humans, and cutting down on energy usage. As chapters in this book demonstrate, the movement has also drawn subsequently ecologists and environmental campaigners, artists, tourists, nightwalkers, and literary champions into its orbit.

Since the discovery and growing awareness of light pollution (Riegel, 1973), professional and amateur astronomers began to work together towards protecting the night sky. The first city light pollution ordinance was conducted in 1958, in the town of Flagstaff, Arizona, home to the Lowell Observatory (Flagstaff Dark Skies Coalition, 2014). As the discovery site for Pluto and dark matter in the universe, the observatory has historical and scientific importance. Flagstaff also housed the Northern Arizona University Astronomic Research Observatory, which drew attention to how light pollution posed a serious threat to future discovery and research. Its supporters, together with advocates for astronomy, recognised the significance of protecting the site from the city's growing light pollution. In response, the ordinance was passed to limit the effects. As the city expanded, the regulations were revised accordingly. Initially prohibiting the use of commercial searchlights, stricter lighting codes were implemented in the 1970s that also banned billboard lighting. Dozens of cities have since followed Flagstaff's example and continue to deter the growth of light pollution.

The first recognised authority on light pollution, the International Dark-Sky Association (International Dark-Sky Association, n.d.), was founded in 1988 by David Crawford, a professional astronomer, and Tim Hunter, a physician and amateur astronomer. A not-for-profit organisation, the International Dark-Sky Association (IDA) has taken the lead on research into the negative impacts of artificial light on human health, wildlife, and climate change. In 2001, the IDA launched the International Dark Sky Places programme (International Dark Sky Places, n.d.) to encourage communities, parks, and protected areas around the world to preserve and protect dark sites through responsible lighting policies and public education. The programme offers five certification categories:

- International dark sky sanctuaries
  Sanctuaries are the most remote, and often the darkest, places in the world whose conservation state is most fragile.
- International dark sky parks
  Parks are publicly or privately owned spaces protected for natural conservation that implement good outdoor lighting and provide dark sky programmes for visitors.
- International dark sky reserves
  Reserves consist of a dark "core" zone surrounded by a populated periphery where policy controls are enacted to protect the darkness of the core.
- Urban night sky places
  Such places are sites near or surrounded by large urban environs whose planning and design actively promote an authentic night-time experience in the midst of significant artificial light at night.
- International dark sky communities
  Communities are legally organised cities and towns that adopt high quality outdoor lighting ordinances and undertake efforts to educate residents about the importance of dark skies.

By January 2023 there were 201 certified dark sky places in the world. These include 115 parks, 38 communities, 20 reserves, 16 sanctuaries, 6 urban night sky places, and 6 dark-sky-friendly developments of distinction. Thus, there are different kinds of designated dark sky places and the crusade for more, including urban areas, continues. For example, Moffat in Dumfries and Galloway, Scotland, was awarded the title of Europe's first dark sky town in 2016, having adopted special street lighting to keep light pollution to a minimum in order to preserve the starry skies. In 2023, to ensure a coherent global identity across its extensive network, the IDA renamed itself DarkSky (2023).

A wider list of dark sky places, including those without official protection, is maintained by the Dark Skies Advisory Group (n.d.) of the International Union for Conservation of Nature. Different countries also have their own initiatives and networks for dark skies. In the UK, for example, the Dark Sky Discovery (n.d.) partnership is a scheme that recognises over 150 areas as dark sky discovery sites, where good quality views of the night sky persist. These are divided into two categories: Milky Way–class and Orion-class sites. Milky Way class means the galaxy must be visible to the naked eye, while Orion class requires that all seven stars in the Orion constellation can be seen unaided. Beyond the international network of DarkSky, there are a number of national efforts such as Commission for Dark Skies (UK) and projects such as Stars4All (EU) that seek to increase public engagement with the issues and advocate for appropriate ways to address the problems caused by light pollution.

The designation of urban dark sky places constitutes an important shift in focus that addresses some of the negative discourses that have construed a division between the rural and the urban, wherein the urban stands as modern in contradistinction to rurality as a threatened historical realm (Dunnett, 2015). In his chapter, Dwayne Avery shows how 19th-century ideas about the moral values of Romanticism, frontier nostalgia and the sublime are inflected with anti-urban and anti-modernist sentiments that continue to inform the contemporary advertising strategies of dark sky tourist marketing. He exemplifies these representations through the retro posters produced by artist Tyler Nordgren that champion the after dark experience of US national parks. This does the dark sky movement a disservice, Avery argues, by construing dark skies as ancient, unchanging, and immune from the polluting tendencies of humans, and yet, as he contends, stars, moon, and planets are now supplemented by numerous satellites that have irrevocably transformed the appearance of the night sky over the past few decades. This needs to be recognised, he maintains, so that astronomical romanticism is historically contextualised by contemporary changes and more mundane encounters with darkness can be recognised as potent.

These urban dark sky places also underline the geographical and environmental diversity of dark sky places, a theme we return to in our concluding chapter. This geographical variety is mirrored by the very different ways of engaging with dark skies that are emerging, as astronomers are joined by botanists, ornithologists, hunters, artists of various inclinations, sound recordists, photographers, tourists, those seeking the therapeutic benefits of darkness, local people, hikers, light designers, storytellers, and writers. The different subjectivities and focus they mobilise testify to different modes of engagement, ways of looking and sensing, scrutinising different objects and experiencing different emotions. In her chapter, such divergencies are brought out by Nona Schulte-Römer from a specifically gendered perspective in an in-depth discussion with four women concerned with dark skies, but from different angles of approach. Occasionally disputatious, these gendered passions and perspectives are themselves revealed to be diverse, sidestepping any essentialist conceits, yet nonetheless, they pose a challenge to the authoritative views of male experts, and especially around the dilemmas of excessive illumination on the grounds of safety. In his chapter, Kerem Asfuroglu draws upon his professional experience of collaborating with dark skies communities to illustrate how the design of lighting schemes can be informed by place-based aesthetic and material qualities that are sympathetic to environmental imperatives and social agents rather than in conflict with them. Using three projects, he explains how the design of more varied, sustainable, and aesthetically attractive nocturnal environments can be evolved through the promotion of the dark sky movement's goals. Asfuroglu's work illustrates how practices can be innovatively applied to rethink and reconfigure the relationship between light and dark.

Sensing dark sky places

While there have been manifold symbolic associations with dark skies, crepuscular settings are not only the focus of cultural meanings but are realms of sensation and affect. While pervasive contemporaneous feelings for those unhabituated to the dark remain shaped by fear and supernatural associations with what cannot be seen, dark sky environments offer a wealth of sensory and affective opportunities for experiencing the world at variance to daytime perception and feeling.

The continuing negative associations of sensations experienced under dark skies belies the notion that our sensory apparatus provides unmediated access to the reality of the world, as we detect sights, sounds, textures, and smells that exist beyond ourselves. However, as Constant Classen (1993, p. 9) points out, "[W]e not only think about our senses, we think through them". Thus, distinctive cultures of looking ensure that sensing with light and dark invariably "involves movement, intention, memory, and imagination" (Macpherson, 2009, p. 1049). We direct attention towards particular objects, interpret the sensory information they provide, and are likely to experience values, fears, likes, and dislikes in accordance with prevailing cultural norms.

Yet despite the ways in which cultural sensory conventions have informed the design and regulation of nocturnal environments, and perhaps because of these ordering processes, the drive for experience in unfamiliar, sometimes overwhelming sensory realms remains powerful. As bodies become accustomed to familiar sensations and smooth progress in predictable spaces, more stimulating engagements with the material world are increasingly sought (Boddy, 1992; Edensor and Millington, 2018). A venture into dark sky places constitutes one such retreat into a space in which unfamiliar sensory experiences may be sought. While a host of recent commercial encounters with darkness are typified by safely regulated environments, which may nonetheless offer potent sensory experiences, a walk through a dark sky park constitutes a less predictable, more variegated, and multi-scalar sensory encounter with darkness and its multiplicities.

To become immersed in a dark sky setting is to move towards an experience of space depicted by Tim Ingold (2011, p. 119) in which the land is not "an interface" separating earth and sky but is a "vaguely defined zone of admixture and intermingling". This is both apprehended visually and through an array of non-visual sensations. This is amply demonstrated in this book by Louise Beer's account of her artistic responses to the night sky, producing ideas and works informed by her New Zealand childhood in which she became habituated to dark skies. Her images are inspired by her enduring and deep connections to the stars, moon, solar system, and universe and the sense of deep historical time that situate the brief, numberless human and non-human lives that have lived and will live on Earth and the astronomical and earthly events that have shaped the landscapes in which they live. This is accompanied by a lingering melancholy for the countless life forms that are rapidly being extinguished in the Anthropocene.

In terms of vision, dominant cultural meanings about landscape remain anchored in romantic notions that posit a detachment between observer and that which is beheld, rendering landscape passive, subject to the onlooker's romantic gaze. At night, the gaze can certainly be arrested by the spectacle of the starry or moon-lit firmament above, but the visual paring down of the land, its monochromatic appearance, provides a dramatic contrast to the variegated colours and complex forms, gradients and distances that characterise its daytime appearance. In a different way to the daylight, the tone, light, darkness, colour, and cloud cover of the "sky in formation" (Ingold, 2011) continuously shifts.

Alphonso Lingis (1998) identifies how changing *levels* of light, characterised by depth of field and brightness, continuously play across the spaces within which we see things and with which we continuously adjust, and this also applies to depths and densities of darkness. This distribution of light and dark across space is shaped by seasonal and weather patterns, diurnal temporalities and longitudinal and latitudinal positions, on the presence of cloud and atmospheric particles. Also, Diane Young describes how colours "animate things in a variety of ways, evoking space, emitting brilliance, endowing things with an aura of energy or light" (2006, p. 173) according to season, time of day, and the light's intensity. Yet at night, human visual perception devolves into a primarily monochromatic experience, characterised by a medley of shades of grey and black, although natural and artificial light can bestow patches of colour on nocturnal landscapes that dramatically stand out from this muted backdrop.

Gloom thus stimulates different ways of looking at landscape and the things in it that stand out. Our perceptual experience of night is also conditioned by the affordances of place, the way moonlight shines on distinctive surfaces or reflects on water, thick shadowy masses of trees and bushes, and silhouetted forms standing out against the sky (Edensor, 2013; Dunn, 2016). Clouds may be tinctured with an orange glow radiated by the lights cast by nearby settlements. Moreover, while the daytime landscape is constituted by a succession of scenes that fade into the distance or come near, and the eye continuously shifts between them, in dark landscapes, vision usually attends to black masses that block light. This is articulated in John Daniel's (2008, p. 23) contention that daylight vision "catches on the surface of things, gets snagged and tugged about by their multiplicity" whereas, he continues, when looking at trees in a forest at night, the sensation is of their "commonality . . . not the names I knew them by but their essential namelessness".

In addition, landscapes seethe with vitality as non-human agencies ceaselessly change them in myriad obvious and unnoticed ways. Landscapes are never pre-formed but are "always in process . . . always in movement, always in making", vitally immanent and emergent (Bender, 2001, p. 3; Benedicktsen and Lund, 2010; Edensor, 2022). Geological, biological, creaturely, chemical, and physical forces combine to transform landscape according to different temporalities. While certain vital elements such as flying creatures, shifting clouds and stars, trees bending in the wind, and looming landforms are visually apprehended at night, the dynamic nocturnal landscape is often more powerfully experienced through emergent smells, sounds and textures. As vision becomes less critical, Tallmadge (2008, p. 140) contends that we become attuned to the landscape through other senses, as the body "relaxes, opens, breathes, extends its attention outward into the world the way a plant feels its way into the soil with roots or into the air with leaves".

In her chapter, Ellen Jeffrey discloses the visual transformation of Finnish woodland into a glowing realm by winter snowfall, and how the landscape is further altered by the sound of her footsteps crunching through snow. In the dark landscape, sound becomes especially accentuated, with the scuttling sounds of small creatures, cascading water, and the wind amongst leaves. Smell too becomes more foregrounded as the scent of pine trees, earth, and fungi permeate the air. Indeed, the qualities of the air, its freshness, moistness, and temperature, command attention.

In recent years, creative non-fiction, especially through the rise of new nature writing, has explored intense, prolonged experiences with landscapes at night, often capturing the potent sensations that arise (Gaw, 2020; Francis, 2019). The sensory experience of rural darkness is beautifully captured in the account by Chris Yates (2012, p. 20) of his nocturnal walks through a hilly area of Southern England. Attuned to the subtle shifts in temperature, flows of air and the scents it carries, he observes how "the evening smells as if the pores of the earth had opened to filter the underground river through all the acres of herb-rich grass around me. After the thin, desiccated atmosphere of the previous week, the humidity seems almost tropical". On a different occasion he remarks how

> as the air calms at the end of a summer's day it begins to generate a subtle and complex interweaving of currents: cool air pouring down from the hill-tops, warm air rising up the valley sides, a localized weather system of minor katabatic [downslope] and anabatic [upslope] winds.
>
> (Ibid., p. 28)

Yates has also become attentive to the sounds of the dark landscape, as he describes in his account of a wood at midnight, which

> had the same stored up warmth and hushed boxed-in quiet as an old high raftered barn—quiet but not soundless. The interior of any summer wood after dark is nearly always gently clicking, rustling, pattering and squeaking . . . sometimes there are unlocatable rufflings as birds fluster their wings in sleep, and if the previous day has been hot there might be the occasional quite sharp crack or creak of cooling dead wood. Even the tiniest mouse-tremble is accentuated in the enclosed, resonant space, which makes a patrolling badger sound like a trundling bear.
>
> (Ibid., p. 70)

During her numerous travels through rural nocturnal Aotearoa, Annette Lees (2022, p. 61) has similarly become highly attuned to the visual elements that grasp her attention, referring to how "landscapes are nuanced with gradations of shadows grey and black—the faint lustre of starlight . . . flakes of moonlight, a flash of bioluminescence, a shimmer of a candle, a spit of firelight, a torch's wobble". She also describes the acoustic identity of places, more evident at night in the absence of visual distractions. For instance, she discerns her location through the sound of rain "as it variously falls on grass, metal, water, rock, sand or leaves" (ibid., p. 65) and depicts how sounds subtly change each hour from 5:00 p.m. to 4:00 a.m. across the night as bird calls rise and fall and the ancient chorus of the cicada spreads across the land. Yet as she disturbingly points out, extensive ecological devastation has undoubtedly reduced the diverse array of non-human sounds, a historic soundscape that can no longer be imagined, and she seeks to identify the more than 50 missing elements that once added to Aotearoa's cacophonous dawn chorus. Such absent sounds include those emitted from the many millions of seabirds that flew in gigantic masses through coastal areas, now only a remnant of these colossal numbers, and the ghosts of the long extinct flightless birds, some giants, that roamed the night.

The sensory experience of darkness, as in all apprehension of landscape, unfolds according to the angle of perception. A journey that includes a walk to the top of an escarpment offers a view backwards to the landscape that has been traversed and a look forward to that which lies ahead. Other views may be typified by a gaze into the middle distance or scrutiny of an unidentifiable object. Yet also critical is how sensations are conditioned by the modes through which we move in dark space, whether as drivers or passengers, walking, running, cycling, sailing, diving, fishing, surfing, or skiing.

Walking in darkness teaches feet to differentiate between textures underfoot, and the body has to anticipate how to walk from moment to moment, with no visual information that prepares the body for a steep ascent or descent (Edensor, 2013). Dancing in the dark is different. In this book, Ellen Jeffrey devises a choreographic practice in a wood at the southern edge of England's Lake District. She is especially drawn to an area in which yew trees create a swathe of canopy that darkens the space and restricts the growth of an understory. The characteristics of the wood ceaselessly shift during the year and through dancing, her body responds to these changes, coming to know this ground kinaesthetically as she moves and is moved by the undulations and materialities of the arboreal environment. As she dances through this dynamic wooded landscape, her imagination seeks to grasp the forms that separate, blur, and merge during the different phases of the night.

Hayden Lorimer (Edensor and Lorimer, 2015, p. 9) details how running at night offers sensory experiences that diverge from running in the daylight. He explains how movements are contingent as "feelings of vulnerability are palpable, borne of not knowing with any real confidence what lies beneath or just ahead. Long strides shorten and stiffen to smaller steps . . . freer movements are pitched off balance by the ever-present worry of putting a foot wrong" as "feelings of fluency deteriorate into a clutter of physical uncertainty". Rather differently, the cyclist is guided along roads and through dark landscapes via the narrow beam of the lamp that cuts an illuminated tunnel through the gloom, highlighting the road surface, trees, and hedges (Cook and Edensor, 2014). In the absence of expansive vistas and multiple scenes, a more intense awareness of the body, its capacities and energies emerge through the changing rhythms of pedalling and physical exertion. Indeed, nocturnal cycling is becoming increasingly popular, as exemplified by the 112-mile Dunwich Run that starts in Hackney, London, and travels through Epping Forest to the Suffolk coastal village of Dunwich (Wills, 2015).

A more insulated experience of travel in the dark is depicted by Sandy Isenstadt (2011, p. 229), who also draws attention to illumination in motion, contending that the early motorist's experience of positioning themselves "at the vertex of a cone of light and propel it across a darkened landscape must count as one of the most startling visual experiences of the twentieth century". At a more rapid rate than cycling, Isenstadt (ibid., p. 218) discusses how car headlights illuminated roadside objects, which "bloomed gradually into nocturnal form, then sharpened for an instant and, just as swiftly were gone". Comparably, on a coach tour that set off to unsuccessfully witness Iceland's aurora borealis, Katrin Lund (2020) discusses how she visually apprehended a host of often unidentifiable forms across the passing nocturnal landscape, while also more intensely experiencing the tactile sensations of mechanised movement through darkness.

The experience of mechanical mobile technological mobilities are also sensorially divergent for passengers on night trains, as evoked by David Bissell (2009, pp. 52–53), who draws attention to "the otherworldly flashes, glows, sparks and orbs that punctuate the dark of night" and "the magical green lights of trackside signals" as the train speeds through nocturnal environments. In a different register, Maria Borovnik (2017) explores how seafarers sensorially experience an ever-shifting environment as their ships move through nocturnal seas, scanning the horizon for landmarks and nautical lights, while developing a sonic attunement to oceanic and mechanical sounds and a tactile sense of the motion engendered by the roiling ocean.

These sensory apprehensions of dark sky settings are accompanied by emotional and affective experiences that potentially solicit creative impulses, imaginative flight, therapeutic possibilities, meditative states and convivial dispositions. Psychologists Anna Steidle and Lioba Werth (2013, p. 76) contend that darkness "elicits a feeling of freedom, self-determination and reduced inhibition . . . and promotes a risky, explorative, and less vigilant task processing style" that fosters creativity. Rob Shaw (2015) explores how darkness offers conditions for meditation and intimacy in the absence of the visible, and Annette Lees (2022, p. 12) concurs, affirming that we are "more primed to experience intimacy, to be softened and receptive to awe and emotion".

Such affective potentialities and multi-sensory attunements were formerly integral to a nocturnal world with little light, a way of living with pervasive darkness that also prompted the acquisition of sensorially informed ways of moving through the night competently. The way ahead could be discerned according to knowledge of the patterns disclosed by stars, and the multiple phases of the moon, the "changing colours and contours in its shape-shifting light" that according to James Attlee (2011, p. 5), are "too subtle for our modern eye to appreciate". Roger Ekirch (2005) also details how parents devised walking exercises to accustom their children to moving through dark space, seeking to develop their aptitude for wayfaring through place via touch, sound, and smell. A range of dark sky experiences are being devised that provide an alternative to daylight sensory experiences, perhaps offering more substantive opportunities for becoming reacquainted with the potent experiences of dark spaces, reattuned to their different sensory pleasures.

Innovations in artistic and tourist engagements at dark sky places

Artistic and creative engagements with dark sky places

The spectacular artificial lighting technologies of the urban night and its relationship with urban darkness has influenced numerous cultural and artistic representations of the nocturnal city in cinema, literature, painting, photography, theatre, and television (Dunn and Edensor, 2020; Sharpe, 2008). This has been recently supplemented by the huge rise in urban light festivals, in which new technological, aesthetic, and conceptual approaches have vastly expanded the range of creative responses to nocturnal urban space, once more re-enchanting the night (Edensor, 2017). These artistic urban innovations have increasingly been complemented by diverse cultural events that have been staged in total or partial darkness. Extraordinary deployments of darkness in ways that solicit changing apprehensions of space, sensory perception, and interactivity have been devised by James Turrell (Sumartojo, 2020) and Tino Seghal (Edensor, 2015), theatrical potential has been elicited by plays that have been staged in the dark (Welton, 2020), and music concerts have also been held to intensify listening in the absence of visual distraction. Darkness has also been inventively exploited in grand immersive cultural experiences staged outside. Notable here are Scottish arts group, NVA, who organised the 2005 event *The Storr: Unfolding Landscape*, which ran across a spectacular mountainside on the Isle of Skye (Morris, 2011), as well as *Speed of Light* in 2012, in which and runners and walkers combined to mark their presence with modest illumination across Edinburgh's volcanic park, Arthur's Seat.

As chapters in this book demonstrate, artistic and creative endeavours are being developed that encourage people to conceive, sense, and affectively engage with dark sky settings in new ways. Some of these projects retain a focus on the visual, others are concerned with attuning visitors to the sonic, aromatic, and tactile sensations that unlit dark places conjure, while some seek to explore cosmic feelings, technological possibilities, and collaborative and interactive initiatives.

Louise Beer's art emerges out of a very personal experience that combines potent memories of the dark skies of her New Zealand childhood, the grief she has suffered following the death of her father and an acute melancholic sense of Earth's diminishing biodiversity. Beer's awareness that everything that constitutes the Earth and the life it contains is derived from the universe adds piquancy to this vision and distinguishes the planet from the potentially lifeless, vast expanse of the solar system in which it spins. She draws out the unique liveliness of the Earth and the human-generated crises that threaten it by adopting various visual strategies. In an audio-video installation, she portrays the spinning Earth over 40 days—an Earth that also produced her father and all our ancestors—as a reassuring vital cycle of change, life, and death that provides constant stability at a time of personal loss. The personal meshes with the cosmic. Another work consists of a manipulated sound piece of nocturnal and dawn birdsong that intensified into a crescendo of grief at the possibility of extinction. In capturing a vision of Earth as an alien planet she created a series of images of white mountains in Tuscany, Italy, that blended photographs sunsets and sunrise, moonlit scenes and night skies that invite the viewer to contemplate the long and slow formation of Earth, far before there were ever eyes to see it . Beer ends her essay with eloquent quotations from astronomers who echo her sentiments about the cosmic preciousness of Earth, invoking the shared, similar responses of scientists and artists.

In exploring the dark sky settings of the Scottish England border, the chapter by Helen McGhie and Natalie Marr provides a compelling insight into the artistic process. The two artist-researchers reflexively identify the evolution of their ideas and the site-specific projects they have undertaken in two dark sky parks through a dialogue. Their often experimental works emerge from engagement with the dark sky sites and through workshops with the people who work in, live in, and care for them: McGhie at Kielder Observatory in north-east England and Marr at Galloway Dark Sky Forest Park in south-west Scotland. Critically, both honour the communities at the sites by registering the values and practices they espouse during different participative exercises. Both seek to foster the emergence of new imaginative, emotional, and sensory experiences. McGhie focuses on offering alternative photographic approaches to the conventional astronomical images produced at dark sky places. She has contingently composed portraits of staff, images of the textures of the ground, photographs converted into sonic forms, and open-air photographic exhibitions staged amongst the trees. Marr has created sound recordings of the nocturnal sounds of the forest, images produced through pinhole photography, and sketches stimulated by her regular encounters with dark skies and sites.

Ysanne Holt's chapter explores how dark skies provide an opportunity for thinking across and beyond the constraints of borders and boundaries and envisioning new forms of connectedness and being together. She focuses on artists who have moved away from the visual and away from structures that may be conceived as having been "imposed" on sites, towards more dematerialised or immaterial art forms and temporary experiences and interactions. Such works sidestep the creation of enduring forms in favour of immanent, unrepeatable, and fleeting experiences. Holt explores the curation of a sensory walk in the dark that offered unexpected experiences of tastes, sounds, and textures and that also made connections with other times and places. She also focuses upon the organising of participative events that explore the potentialities of the situated, sonic experience of off-grid, dark sky landscapes to stoke environmental consciousness and a sense of place. Such work seeks to forge new connections between people through collective participation in using alternative forms of digital technology that lie outside mainstream networks.

Dark sky tourism

Darkness has long been economically deployed to entice visitors to a wealth of attractions. As Alice Barnaby (2020) details, as modern illumination spread, during the first half of the 19th century, the Hermit's Grotto, Submarine Caves, and Dark Walk in the Vauxhall Pleasure Gardens, London, produced a recreational space in which visitors could experience depths of light and darkness. These enticements have been superseded by ghost trains and thrill rides that have also relied on the potential of darkness to enhance experience. Once more, this underlines how experience of darkness provides a corrective for ocularcentrism, confirming that "tourists encounter the world multisensually and multidimensionally" (Scarles, 2009, p. 467), through tactility, sound, smell, and taste.

In recent decades, the expansion of tourism as an economic imperative and cultural practice has incorporated an increasingly diverse range of attractions, activities and destinations across the world (Urry, 1990). This extension reaches into the night, with the nocturnalisation of tourism evident in the growing popularity of visits to museums in the dark, graveyard tours, ghost walks, and bat walks (Eldridge and Smith, 2019). Celestial tourism is also growing (Weaver, 2011). Tours to the Arctic Circle are growing in popularity as visitors travel to experience the "Land of the Midnight Sun" (Birkeland, 1999), while other tourists venture to Iceland, northern Scandinavia, and Canada to experience the aurora borealis (Lund, 2020). Chapters in this book reveal this development of a range of tourist experiences in dark sky places, and while many tourists are motivated by astronomical spectacles, other diverse pleasures are being adopted to attract visitors.

Importantly, MacMillan *et al.*'s chapter in this book foregrounds two essential dimensions in developing tourism in dark sky places. First, they emphasise the importance of the advocacy and participation of the local community in campaigning for dark sky designation and subsequently, in devising and participating in key events designed for both residents and tourists. Second, they show that while astrotourism may continue to attract most visitors to Mayo Dark Sky Park, a medley of storytelling and literary festivals, nature walks, the staging of historical events, and the aesthetic pleasures of innovative lighting schemes offer alternative delights for both tourists and locals. In her chapter, Neha Khetrapal takes a different perspective in advocating how dark skies are inextricably entangled with ancient celestial myths that are materialised in the sculptural and material designs of significant heritage sites, such as at the central Indian temple complex of Khajuraho (also see Khetrapal and Bhatia, 2022).

Conclusion

In this introduction we have sought to provide a substantive context for the chapters that follow by highlighting the histories of dark skies, multiple meanings of darkness, the loss of darkness due to over-illumination, the rise of the dark sky movement, and how creative engagements with dark sky places offer new insights for reconnecting with dark skies. We conceive of this book as contributing to a conversation that is multi-dimensional, multi-disciplinary, and continually emerging. We contend that the chapters in this volume reveal the sheer diversity of debates, experiences, historical contexts, and geographical settings associated with dark skies. We further explore the implications of the issues they raise in the concluding chapter.

References

Attlee, J. (2011) *Nocturne: A journey in search of moonlight*. London: Hamish Hamilton.

Barnaby, A. (2020) 'In the night garden: Vauxhall pleasure gardens 1800–1859', in Dunn, N. and Edensor, T. (eds.) *Rethinking darkness: Cultures, histories, practices*. London: Routledge, pp. 50–60.

Baudrillard, J. (1986) *America*. London: Verso.

Bender, B. (2001) 'Introduction', in Bender, B. and Winer, M. (eds.) *Contested landscapes: Movement, exile and place*. Oxford: Berg.

Benedicktsen, K. and Lund, K. (2010) 'Introduction: Starting a conversation with landscape', in Benedicktsen, K. and Lund, K. (eds.) *Conversations with landscape*. Farnham: Ashgate.

Bille, M. and Sørensen, T. (2007) 'An anthropology of luminosity: The agency of light', *Journal of Material Culture*, 12(3), pp. 263–284. http://dx.doi.org/10.1177/1359183507081894.

Birkeland, I. (1999) 'The Mytho-poetic in Northern travel', in Crouch, D. (ed.) *Leisure/tourism geographies*. London: Routledge, pp. 17–33.

Bissell, D. (2009) 'Visualising everyday geographies: Practices of vision through travel-time', *Transactions of the Institute of British Geographers*, 34(1), pp. 42–60.

Boddy, T. (1992) 'Underground and overhead: Building the analogous city', in Sorkin, M. (ed.) *Variations on a theme park*. New York: Hill and Wang.

Bordin, G. (2020) 'Inuit's perception of darkness: A singular feature', in Dunn, N. and Edensor, T. (eds.) *Rethinking darkness: Cultures, histories, practices*. London: Routledge, pp. 91–103.

Borovnik, M. (2017) 'Nighttime navigating: Moving a container ship through darkness', *Transfers*, 7(3), pp. 38–55.

Boutsikas, E. (2017) *The role of darkness in ancient Greek religion and religious practice*. Oxford: Oxford University Press.

Brands, J., Schwanen, T. and Van Aalst, I. (2015) 'Fear of crime and affective ambiguities in the night-time economy', *Urban Studies*, 52(3), pp. 439–455.

Brox, J. (2010) *Brilliant: The evolution of artificial light*. New York, NY: Houghton Mifflin Harcourt.

Chepesiuk, R. (2009) 'Missing the dark: Health effects of light pollution', *Environmental Health Perspectives*, 117(1), pp. A20–A27. https://doi.org/10.1289/ehp.117-a20.

Cinzano, P., Falchi, F. and Elvidge C.D. (2001) 'The first World Atlas of the artificial night sky brightness', *Monthly Notices of the Royal Astronomical Society*, 328(3), pp. 689–707. https://doi.org/10.1046/j.1365-8711.2001.04882.x.

Clarke, S. (2020) *Beneath the night: How the stars have shaped the history of humankind*. London: Guardian Faber Publishing.

Classen, C. (1993) *Worlds of sense: Exploring the senses in history and across cultures*. London: Routledge.

Collins, J. and Jervis, J. (2008) 'Introduction', in Collins, J. and Jervis, J. (eds.) *Uncanny modernity: Cultural theories, modern anxieties*. London: Palgrave Macmillan.

Cook, M. and Edensor, T. (2014) 'Cycling through dark space: Apprehending landscape otherwise', *Mobilities*, 12(1), pp. 1–19.

Daniel, J. (2008) 'In praise of darkness', in Bogard, P. (ed.) *Let there be night: Testimony on behalf of the dark*. Reno, NV: University of Nevada Press.

Dark Skies Advisory Group. (n.d.) Available at: http://darkskyparks.org/dark-skies-and-nature-conservation/.

Dark Sky Discovery. (n.d.) Available at: www.darkskydiscovery.org.uk.

DarkSky. (2023) *DarkSky*. Available at: http://darksky.org/

Davies, T., Duffy, J., Bennie, J. and Gaston, K. (2015) 'Stemming the tide of light pollution encroaching into marine protected areas', *Conservation Letters*, 9(3), pp. 151–236. https://doi.org/10.1111/conl.12191.

Davies, T. and Smyth, T. (2018) 'Why artificial light at night should be a focus for global change research in the 21st century', *Global Change Biology*, 24(3), pp. 872–882. https://doi.org/10.1111/gcb.13927.

Drake, N. (2019) 'Our nights are getting brighter, and earth is paying the price', *National Geographic*, 3 April. Available at: www.nationalgeographic.com/science/2019/04/nights-are-getting-brighter-earth-paying-the-price-light-pollution-dark-skies/.

Dunn, N. (2016) *Dark matters: A manifesto for the nocturnal city*. New York, NY: Zero Books.

Dunn, N. and Edensor, T. (eds.) (2020) *Rethinking darkness: Cultures, histories, practices*. London: Routledge.

Dunnett, O. (2015) 'Contested landscapes: The moral geographies of light pollution in Britain', *Cultural Geographies*, 22(4), pp. 619–636. https://doi.org/10.1177/1474474014542.

Ebbensgaard, C. and Edensor, T. (2021) 'Walking with light and the discontinuous experience of urban change', *Transactions of the Institute of British Geographers*, 46, pp. 378–391. https://doi.org/10.1111/tran.12424.

Edensor, T. (2013) 'Reconnecting with darkness: Experiencing landscapes and sites of gloom', *Social and Cultural Geography*, 14(4), pp. 446–465. https://doi.org/10.1080/14649365.2013.790992.

Edensor, T. (2015) 'Light art, perception, and sensation', *The Senses and Society*, 10(2), pp. 138–157.

Edensor, T. (2017) *From light to dark. Daylight, illumination, and gloom*. Minneapolis, MN: University of Minnesota Press.

Edensor, T. (2022) *Landscape, materiality and heritage: An object biography*. London: Palgrave Macmillan.

Edensor, T. and Lorimer, H. (2015) 'Landscapism at the *speed of light*: Darkness and illumination in motion', *Geografiska Annaler: Series B, Human Geography*, 97(1), pp. 1–16.

Edensor, T. and Millington, S. (2018) 'Learning from Blackpool promenade: Re-enchanting sterile streets', *The Sociological Review*, 66(5), pp. 1017–1035.

Ekirch, R. (2005) *At day's close: Night in times past*. London: W.W. Norton and Company.

Eldridge, A. and Smith, A. (2019) 'Tourism and the night: Towards a broader understanding of nocturnal city destinations', *Journal of Policy Research in Tourism, Leisure and Events*, 11(3), pp. 371–379. https://doi.org/10.1080/19407963.2019.1631519.

Falchi, F., Cinzano, P., Duriscoe, D., Kyba, C. C. M., Elvidge, C. D., Baugh, K., Portnov, B. A., Rybnikova, N. A. and Furgoni, R. (2016) 'The new world atlas of artificial night sky brightness', *Science Advances*, 2(6), e1600377. https://doi.org/10.1126/sciadv.1600377.

Falchi, F., Cinzano, P., Elvidge, C., Keith, D. M. and Haim, A. (2011) 'Limiting the impact of light pollution on human health, environment and stellar visibility', *Journal of Environmental Management*, 92(10), pp. 2714–2722.

Flagstaff Dark Skies Coalition. (2014) Available at: https://flagstaffdarkskies.org/international-dark-sky-city/flagstaffs-battle-for-dark-skies/.

Foster, J., Smolka, J., Nilsson, D. and Dacke, M. (2018) 'How animals follow the stars', *Proceedings of the Royal Society B*, 285. http://doi.org/10.1098/rspb.2017.232220172322.

Francis, T. (2019) *Dark skies: A journey into the wild night*. London: Bloomsbury.

Galinier, J., Becquelin, A., Bordin, G., Fontaine, L., Fourmaux, F., Ponce, J., Salzarulo, P., Simonnot, P., Therrien, M. and Zilli, I. (2010) 'Anthropology of the night: Cross-disciplinary investigations', *Current Anthropology*, 51(6), pp. 819–847.

Gallan, B. and Gibson, C. (2011) 'Commentary: New dawn or new dusk? Beyond the binary of night and day', *Environment and Planning A*, 43, pp. 2509–2515.

Gallaway, T. (2015) 'The value of the night sky', in Meier, J., Hasenöhrl, U., Krause, K. and Pottharst, M. (eds.) *Urban lighting, light pollution and society*. London: Routledge.

Gaston, K., Visser, M. and Hölker, F. (eds.) (2015) 'The biological impacts of artificial light at night: From molecules to communities', *Philosophical Transactions of the Royal Society B*, 370, p. 166.

Gaw, M. (2020) *Under the stars: A journey into light*. London: Elliott & Thompson.

Gonlin, N. and Nowell, A. (eds.) (2018) *Archaeology of the night: Life after dark in the ancient world*. Boulder: University Press of Colorado.

Green, J., Perkins, C., Steinbach, R. and Edwards, P. (2015) 'Reduced street lighting at night and health: A rapid appraisal of public views in England and Wales', *Health and Place*, 34, pp. 171–180. https://doi.org/10.1016/j.healthplace.2015.05.011.

Haim, A., Scantlebury, D. and Zubidat, A. (2019) 'The loss of ecosystem-services emerging from artificial light at night', *Chronobiology International*, 36(2), pp. 296–298. https://doi.org/10.1080/07420528.2018.1534122.

Henderson, D. (2010) 'Valuing the stars: On the economics of light pollution', *Environmental Philosophy*, 7(2), pp. 17–26.

Hutton, R. (2018) 'The strange history of British archaeoastronomy', in Gunzburg, D. (ed.) *The imagined sky: Cultural perspectives*. Sheffield: Equinox, pp. 1–22.

Ingold, T. (2011) *Being alive: Essays on movement, knowledge and description*. London and New York: Routledge.

International Dark-Sky Association. (n.d.) Available at: www.darksky.org/.

International Dark Sky Places. (n.d.) Available at: www.darksky.org/our-work/conservation/idsp/.

Isenstadt, S. (2011) 'Auto-specularity: Driving through the American night', *Modernism/Modernity*, 18(2), pp. 213–231.

Jenkins, R. (2000) 'Disenchantment, enchantment and re-enchantment: Max Weber at the millennium', *Max Weber Studies*, 1(1), pp. 11–32.

Jung, C. (1976) 'Letter to Sigmund Freud, 8 May 1911', in Adler, G., Jung, C. G., Hull, R. F. C. and Aniela, J. (eds.) *Letters of C. G. Jung, Volume 2, 1951–1961*. London: Routledge and Kegan Paul, pp. 463–464.

Khetrapal, N. and Bhatia, D. (2022) 'Our brightly-lit future: Exploring the potential for astrotourism in Khajuraho (India)', *The Canadian Geographer*, 66(3), pp. 621–627.

Koslofsky, C. (2011) *Evening's empire: A history of night in early modern Europe*. Cambridge: Cambridge University Press.

Kumar, A. and Shaw, R. (2019) 'Transforming rural light and dark under planetary urbanisation: Comparing ordinary countrysides in India and the UK', *Transactions of the Institute of British Geographers*, 45(1), pp. 155–167.

Kyba, C., Hänel, A. and Hölker., F. (2014) 'Redefining efficiency for outdoor lighting', *Energy & Environmental Science*, 7, pp. 1806–1809.

Lees, A. (2022) *After dark: Walking into the nights of Aotearoa*. Potton and Burton: Nelson.

Levin, A. (2019) *Incandescent: We need to talk about light*. Salford: Saraband.

Lingis, A. (1998) *Foreign bodies*. Bloomington, IN: Indiana University Press.

Lund, K. (2020) 'Creatures of the night: Bodies, rhythms and Aurora Borealis', in Dunn, N. and Edensor, T. (eds.) *Rethinking darkness: Cultures, histories, practices*. London: Routledge, pp. 127–137.

MacPherson, H. (2009) 'The intercorporeal emergence of landscape: Negotiating sight, blindness, and ideas of landscape in the British countryside', *Environment and Planning A*, 41, pp. 1042–1054.

Marshack, A. (1972) *The roots of civilization: The cognitive beginnings of man's first art, symbol and notation*. New York, NY: McGraw-Hill.

McCafferty, P. (2018) 'Comets and meteors: The ignored explanations for myths and apocalypse', in Ginzburg, D. (ed.) *The imagined sky: Cultural perspectives*. Sheffield: Equinox, pp. 23–43.

McQuire, S. (2008) *The media city: Media, architecture and urban space*. London: Sage.

Meier, J., Hasenöhrl, U., Krause, K. and Pottharst, M. (eds.) (2014) *Urban lighting, light pollution and society*. New York, NY: Taylor & Francis.

Melbin, M. (1978) 'Night as frontier', *American Sociological Review*, 43(1), pp. 3–22.

Morris, N. (2011) 'Night walking: Darkness and sensory perception in a night-time landscape installation', *Cultural Geographies*, 18(3), pp. 315–342. https://doi.org/10.1177/1474474011410277.

Nasaw, D. (1999) *Going out: The rise and fall of public amusements*. Boston: Harvard University Press.

Noon, K. and De Napoli, K. (2022) *Astronomy: Sky country*. Melbourne: Thames and Hudson.

Nowell, A. and Gonlin, N. (2020) 'Affordances of the night: Work after dark in the ancient world', in Dunn, N. and Edensor, T. (eds.) *Rethinking darkness: Cultures, histories, practices*. London: Routledge, pp. 27–37.

Otter, C. (2008) *The Victorian eye: A political history of light and vision in Britain, 1800–1910*. Chicago, IL: University of Chicago Press.

Palmer, B. (2000) *Cultures of darkness: Night travels in the histories of transgression: From medieval to modern*. New York, NY: Monthly Review Press.

Pawson, S. and Bader, M. (2014) 'LED lighting increases the ecological impact of light pollution irrespective of color temperature', *Ecological Applications*, 24(7), pp. 1561–1568. https://doi.org/10.1890/14-0468.1.

Pritchard, S. (2017) 'The trouble with darkness: NASA's Suomi satellite images of earth at night', *Environmental History*, 22(2), pp. 312–330.

Pyle, R. (1978) 'The extinction of experience', *Horticulture*, 56, pp. 64–67.

Rich, C. and Longcore, T. (eds.) (2006) *Ecological consequences of artificial night lighting*. Washington, DC: Island Press.

Riegel, K. (1973) 'Light pollution: Outdoor lighting is a growing threat to astronomy', *Science*, 179, pp. 1285–1291. https://doi.org/10.1126/science.179.4080.1285.

Rijo-Ferreira, F. and Takahashi, J. (2019) 'Genomics of circadian rhythms in health and disease', *Genome Med*, 11(82), pp. 1–6. https://doi.org/10.1186/s13073-019-0704-0.

Sánchez-Bayo, F. and Wyckhuys, K. (2019) 'Worldwide decline of the entomofauna: A review of its drivers', *Biological Conservation*, 232, pp. 8–27. https://doi.org/10.1016/j.biocon.2019.01.020.

Savela, T (2023) 'Like night and day: Channelling desires through landscapes and nightscapes', *Landscape Research*, 48(4), pp. 544–560. https://doi.org/10.1080/01426397.2023.2167962.

Scarles, C. (2009) 'Becoming tourist: Renegotiating the visual in the tourist experience', *Environment and Planning D: Society and Space*, 27(3), pp. 465–488.

Schivelbusch, W. (1988) *Disenchanted night: The industrialisation of light in the nineteenth century*. Oxford: Berg.

Schlör, J. (1998) *Nights in the big city*. London: Reaktion.

Sharpe, W. (2008) *New York nocturne: The city after dark in literature, painting, and photography*. Princeton, NJ: Princeton University Press.

Shaw, R. (2015) 'Controlling darkness: Self, dark and the domestic night', *Cultural Geographies*, 22(4), pp. 585–600. https://doi.org/10.1177/147447401453925.

Sloane, M. (2016) 'Darkness is a luxury not granted to Britain's council estates', *The Guardian*, 19 May. Available at: www.theguardian.com/housing-network/2016/may/19/darkness-luxury-britains-council-estates-lighting-cctv.

Smyth, T., Wright, A., McKee, D., Tidau, S., Tamir, R., Dubinsky, Z., Iluz, D. and Davies, T. (2021) 'A global atlas of artificial light at night under the sea', *Elementa: Science of the Anthropocene*, 9(1), p. 00049. https://doi.org/10.1525/elementa.2021.00049.

Soga, M. and Gaston, K. J. (2016) 'Extinction of experience: The loss of human-nature interactions', *Frontiers in Ecology and the Environment*, 14, pp. 94–101. https://doi.org/10.1002/fee.1225.

Steidle, A. and Werth, L. (2013) 'Freedom from constraints: Darkness and dim illumination promote creativity', *Journal of Environmental Psychology*, 35(1), pp. 67–80.

Sumartojo, S. (2020) 'On darkness, duration and possibility', in Dunn, N. and Edensor, T. (eds.) *Rethinking darkness: Cultures, histories, practices*. London: Routledge, pp. 192–201.

Sutton, P. and Elvidge, C. (2015) 'Night lights: An indicator of the good life?' in Meier, J., Hasenöhrl, U., Krause, K. and Pottharst, M. (eds.) *Urban lighting, light pollution, and society*. London: Routledge, pp. 284–298.

Tallmadge, J. (2008) 'Night vision', in Bogard, P. (ed.) *Let there be night: Testimony on behalf of the dark*. Reno, NV: University of Nevada Press.

Urry, J. (1990) *The tourist gaze: Leisure and travel in contemporary societies*. London: Sage.

Van Doren, B., Horton, K., Dokter, A., Klinck, H., Elbin, S. and Farnsworth, A. (2017) 'Intense urban lights alter bird migration', *Proceedings of the National Academy of Sciences*, 114(42), pp. 11175–11180. https://doi.org/10.1073/pnas.1708574114.

Weaver, D. (2011) 'Celestial ecotourism: New horizons in nature-based tourism', *Journal of Ecotourism*, 10(1), pp. 38–45. https://doi.org/10.1080/14724040903576116.

Welton, M. (2020) 'Going dark: The theatrical legacy of Battersea arts Centre's playing in the dark season', in Dunn, N. and Edensor, T. (eds.) *Rethinking darkness: Cultures, histories, practices*. London: Routledge, pp. 179–191.

Williams, R. (2008) 'Nightspaces: Darkness, deterritorialisation and social control', *Space and Culture*, 11(4), pp. 514–532. https://doi.org/10.1177/120633120832011.

Wills, D. (2015) *At night: A journey round Britain from dusk to dawn*. Basingstoke: AA Publishing.

Yates, C. (2012) *Nightwalk: A journey to the heart of nature*. London: Collins.

Young, D. (2006) 'The colours of things', in Spyer, P., Tilley, C., Kuechler, S. and Keane, W. (eds.) *The handbook of material culture*. London: Sage.

Part 2

Creative engagements with dark places

2 Creative approaches to dark skies research

A dialogue between two artist-researchers

Helen McGhie and Natalie Marr

Introduction

Though the arts are vital in engaging our imaginations of the night sky and are increasingly featured in the public programming of international dark sky parks (IDSPs), the role and value of arts-based approaches to dark sky research remains underexplored. Our collaborative chapter addresses this through a dialogue in which we, two artist-researchers, share experiences, insights, and challenges from our research, undertaken in partnership with two UK-based IDSP contexts, Kielder Observatory Astronomical Society (KOAS) in Northumberland International Dark Sky Park, England and Galloway Forest Dark Sky Park (GFDSP), Scotland. We introduce our individual projects before situating our work at the intersection between a renewed cultural engagement with dark skies and broader commitments to develop interdisciplinary research into dark skies. While the arts are considered valuable tools for communicating the wonders of the night sky and for facilitating meaningful relationships with dark landscapes, far less attention has been afforded to the value of the artistic process itself, which we will consider here. As artist-researchers working in and with IDSPs over several years, we are uniquely positioned to share the value of creative research partnerships with IDSPs and how they can enable stakeholders to reflect on and contribute to the dark sky experience. We do so through a dialogue that explores synergies between our respective projects—one based in the field of photography, the other in cultural geography—sharing our experiences, insights, and challenges of working with IDSPs as research sites and partners. Reflective in style and concerned with questions of method, site, and practice, our dialogue demonstrates the receptive nature of creative research and how its production can contribute to IDSP activities (for the public and internal stakeholders) to enrich existing public events and offer communities renewed contemplations with the environment. Artist-researchers use creative methods to learn from and respond to places and people; they can enrich cultural narratives around darkness and the night sky, meeting the aspirations and challenges of the increasingly interdisciplinary field of night studies. We close the chapter with a set of critical reflections on our contributions and share propositions to expand our work. These will consider the value of situating artist-researchers in

DOI: 10.4324/9781003408444-4
This chapter has been made available under a CC-BY-NC-ND 4.0 license

IDSPs, demonstrate how there is further room for artist-researchers to explore, and suggest future opportunities for opening discussions around this working approach.

Arts-based approaches to dark skies

IDSPs are places that provide public access to pristine dark skies not only through specialist facilities and knowledge but also through recreational and cultural activities that seek to integrate "all that lies above—sights, sounds, hopes . . . into our total lives" (Jafari, 2007, p. 55). IDSP public programmes increasingly feature the input of creative practitioners whose contributions include permanent sculptures (Kielder Art and Architecture, 2023), artist residencies (National Park Service, 2022; King, 2022), and temporary artworks, workshops, and events as part of dark sky festivals (An Lanntair, 2022; Mayo Dark Sky Festival, n.d.). Creative contributions are considered valuable for communicating the universe's wonders through stunning visuals and expressive media and for deepening public relationships with the night sky through installations that inspire and enchant (Charlier, 2018; Jerram, n.d.). While this is heartening to us as artist-researchers, attention is often given to the public-facing outcomes of artworks rather than the process of artistic production itself, including planning, practice, and reflection.

Our chapter addresses this through a reflective dialogue that draws on direct experience and contemplates the production of research—our thought processes, stakeholder engagements, and site relations. From different perspectives, our projects share a commitment to making work in dialogue with IDSPs, using arts-based methods to cultivate new perspectives of, and relationships with, the night sky and to contribute to the cultural development of dark sky places. Arts-based methods position artworks (paintings, installations, photographs) and creative participation (conversations, workshops, happenings) as generative sites of research and exchange (Candy, 2019; Barrett and Bolt, 2010). Oriented to "the collective creative potential of a given constituency or site" (Kester, 2004, p. 24; see also, Hawkins, 2015), arts-based methods de-centre the artist as primary meaning-maker and facilitate the mutual exploration of ideas and processes to produce new and unexpected findings.

With artist training from the Royal College of Art (London, UK), Helen brings her creative expertise in using "staged" photography (Cotton, 2020, pp. 38–69) to articulate the experiences of astronomers at KOAS. Helen's research—*Close Encounters: Developing Photographic Outcomes in Collaboration with a Science Communication Organisation in Northumberland International Dark Sky Park*—explores the value of an artist embedded within a small-scale science outreach charity. It uses methods of arts-based research and creative reflective practice to engage with the dark sky community during an extended five-year research residency. Helen learned the personal stories of observers, such as why they observe and what inspires them about dark skies. In response, she created a series of photographic outputs: on-location portraits (of staff, volunteers, and visitors), otherworldly landscapes, still-life photography, and a short film. Images were informed by conversations with the people of KOAS, which she visually interpreted before later disseminating as artworks in a public museum context and *in situ* at KOAS, during observing sessions and special "arts" events, which included an online exhibition (embedded within the KOAS website), an immersive sound trail, and an outdoor photography exhibition. Creative production and dissemination generated opportunities to engage the community through impromptu chats during observatory events and during structured art experiences led by Helen.

Natalie's research—*Windows to the Universe: Mapping the Values of the Galloway Forest Dark Sky Park*—explores the sociocultural values of dark skies from the situated perspective of the GFDSP, which was awarded dark sky status by the International Dark-Sky Association in 2009 and is the first of its kind in the UK and Europe. The project maps how the park and its values are variously imagined, experienced, and enacted by stakeholders in the years following international designation. Currently situated in cultural geography and trained in contemporary fine art practice, film-making, and experimental media, Natalie's research practice is interdisciplinary, composed of qualitative research methods, ethnography, and creative enquiry through long-exposure photography, audio recording, and embodied, participative practice. Interested in how dark sky values emerge *in situ*, her research has increasingly attended to the aesthetic experience of the dark sky park and the various activities and practices that shape it. From public stargazing events and guided tours to informal gatherings of local residents and contingent encounters with other nocturnal inhabitants, the project presents a rich ethnography of the social lives and landscapes of the GFDSP and elaborates an expanded vocabulary of dark sky practice. Conducted during a period in which the GFDSP celebrated its ten-year anniversary, *Windows to the Universe* engages stakeholders in collective reflection, contributing to the park's future development.

Our respective projects, while different in form and discipline—one a PhD by practice based in the field of photography, the other a PhD by thesis based in cultural geography—involve external partners in the research design and delivery. Such research containers allow us, as artists, to explore the possibilities of arts-based research in dark sky contexts over a sustained period and in conversation with stakeholders. This chapter brings our respective projects into a conversation to share practice, critically reflect on the impact of our work, and identify opportunities for future partnerships between artist-researchers and IDSPs.

A dialogue

Our process began with a few initial questions shared in an online document, which grew into a more fluid correspondence, as each response prompted a new question. Our dialogue is inspired by discussions we had at *Dark Sky Meeting*, an interdisciplinary conference hosted by Exeter College, Oxford, where we met for the first time in early 2022. This encouraged us to explore the intersections of our work, the value of creative practice under dark skies, and the importance of building stronger connections between researchers and artists in this growing field of study.

Dialogue can be an open-ended and reflective mode of investigation that shifts focus from the fashioning of distinct outputs or findings to the fostering of participative and site-responsive modes of knowledge production (Kester, 2004, p. 24). As a lively practice of speaking, listening, and responding, it fosters a generative space for exploration and speculation, allowing unexpected questions and solutions to emerge (Edwards, Collins and Goto, 2016). We employ dialogue here as a critical tool to open discussions around the value of creative practice in dark sky contexts, share practice, and identify new trajectories for our work at the intersection of academic research, creative production, and dark sky practice.

Natalie (N): What were the origins of your project, and what were your initial questions or points of reference?

Helen (H): I have printed my photographs in the darkroom for years; my process involves intuitively selecting images, testing exposures, and considering how my images communicate to the world. For me, darkness is not only a practical necessity for analogue photography, but a productive space to imagine and create stories, where the quiet moments between exposures enable contemplation. Seeking inspiration for darkness in a new context, I applied for a PhD opportunity in partnership with KOAS, located in the second largest IDSP in Europe; the project sought mutual benefit for an artist and the observatory when they worked together. At the start of the project, KOAS mainly used amateur astrophotography for marketing and print sales, or open-source space telescope images in presentations. Artists interested in astronomy practised astrophotography or worked with available astro-images to communicate tensions between "art and science" ask political questions or explore personal narratives. By producing photography, I wished to respond to my experiences of darkness, informed by the dark sky communities and their situated experiences of night (Figures 2.1 and 2.2).

Whilst seeking creative inspiration within darkness, I wanted to explore KOAS as a small-scale astronomy charity, with a community passionate about dark skies, whilst discovering new methods for artists to engage in partnership working. Whilst there is a tradition of creative practitioners undertaking short residencies at local observatories (King, 2022; Rickett, 2014) and at large institutions (ESA, 2023; Arts at CERN, 2023), artists are understood to benefit more than the hosts (Glinkowski and Bamford, 2009), and there is a need to explore if more mutual benefits can be gained for both artists and organisations over a longer term. As IDSPs are in publicly accessible rural areas (if one can drive), protected for "scientific, natural, educational, and/or cultural heritage resources, and/or for public enjoyment" (International Dark-Sky Association, 2018, p. 3), there is the scope for artists to position their creative expertise as an IDSP "resource" whilst finding inspiration under dark skies. At the start of my project, I learned that there was limited access to KOAS staff and space during busy public events;

Figure 2.1 "Wanderer, 16.02.20", portrait of a KOAS astronomer (2020). Image credit: Helen McGhie.

my early visits could only scratch the surface of possibility. Undertaking an *extended* residency for five years may deepen my relationships with staff, and my understanding of the observatory and Kielder's darkness, whilst testing different photographic responses. Through my photographic encounters with dark skies (and reflections in the darkroom afterwards), I hoped to develop a new visual language of *being* in an IDSP.

N: My relationship with the GFDSP began during my master's degree in filmmaking, for which my final project was a short film that explored stargazing, darkness, and grief in conversation with family and friends, and amateur astronomers and dark sky rangers I had met in Galloway. I saw the PhD project advertised and was compelled by its framing of the dark sky park as a generative site for creative, interdisciplinary

Figure 2.2   "Dark Adaptation", landscape photograph at KOAS (2019). Image credit: Helen McGhie.

research. My research aimed to investigate the cultural and social values of the GFDSP, the phenomenon of stargazing, and the impact of international dark sky designation on the region. I wanted to explore the intangible values of dark skies—what it *feels* like to look at the stars and experience natural darkness, how the dark sky park features in people's lifeworlds, and what motivates stakeholders to become involved. My approach was informed by cultural geography and environmental humanities literature with a focus on light, darkness, and the night sky as underexplored features of landscape, as well as non-representational approaches that attend to the experiential, sensory, and situated dimensions of environmental relation (Sumartojo and Pink, 2019; Vannini, 2015).

I was also fascinated by the tension between international designation as a narrative "from above" and the actual practice of stewarding a dark sky park "from below". How were the core values of the international dark sky movement mobilised and enacted by the GFDSP and what other values were emerging or already present? Ada Blair's (2016) research on the Dark Sky Island of Sark was a key reference. Blair shared stories of community organising and relationships with the stars and darkness which were developing through informal stargazing and nightwalking practices. It presented an account of an IDSP as not just visited but inhabited and practised. Like Blair, I was interested in what it's like to live and work there. What motivates such an endeavour, and what sustains it? Who is involved, what does it mean to them, and does its meaning change over time?

H: Was there an existing context for "art" when we arrived at our respective IDSPs, and if so, did it shape our approach?

N: The GFDSP's Recreation Services Manager Keith (my main point of contact) was keen to explore Galloway's dark skies beyond astronomy through social and cultural events that incorporated the arts. The research project was one vehicle through which to do this. My fieldwork began with two events that were already taking place in the dark sky park. The first was the inaugural European Dark Sky Places Conference (EDSP), organised by the Galloway and Southern Ayrshire Biosphere, Forest Enterprise Scotland, and the International Dark-Sky Association, which featured a panel on creative approaches to darkness and light. The second event was *Sanctuary*, a 24-hour experimental arts festival that had been running biannually since 2013 in one of the darkest parts of the park. I took part in both events as a researcher speaking on the conference panel and as an artist commissioned to make a site-specific artwork at the festival. Occupying this dual role of artist and researcher encouraged me to combine research methods throughout my fieldwork (for example, using sensory ethnography in interviews or giving participants my handmade cameras with which to make their own images). It also prompted me to reflect on how specific contexts and communities of practice (a specialist conference vs. a public arts festival) shaped what was possible in the research. At the conference, I presented my research as part of a community of dark sky researchers and practitioners seeking to exchange ideas and develop solutions to the challenges of artificial light pollution. In contrast, at *Sanctuary*, the slow and receptive qualities of the artwork—an installation where visitors could listen to the sounds of the forest at night through sensitive microphones and a bat detector—instilled a sense of my research as emergent, participative, and site-specific.

H: KOAS is part of Kielder's Art and Architecture Trail, designed as an architectural installation for visitors to engage with during the day and at night. Considered as a "transgressive observatory" by Peter Sharpe, Kielder's art curator (McGhie, 2020), its unusual design was completed by Charles Barclay Architects in 2008 to look like the deck of a ship emerging from the landscape, reminding visitors of astronomy's relationship with nautical navigation. There was a process of creative "pre-visualisation" during its design phase, where stakeholders imagined its appearance, visitor engagement and relation to artworks nearby, notably James Turrell's Skyspace (Visit Kielder, 2000). Artistic responses were also invited for the launch, where Alec Finlay realised a poetry installation for the windmill (2008a) and the *One Hundred Year Star Diary* publication (2008b). Despite its artistic grounding, KOAS developed as a science attraction and amateur astrophotography was the main creative practice when I arrived in 2017.

I sought to provoke the practice of photography at KOAS, to test out alternative modes of image-making, of capturing one's *encounters* with and emotions under darkness. Recognising the practice of "pre-visualisation" during KOAS's design (Kielder Observatory, 2021), I considered parallels with the imaginative process of making analogue photographs unaided by a digital preview screen; one pictures light, contrast, and composition in their mind's eye (Adams, 1948), not knowing the results until the film is processed. I initially captured staff portraits using flash bursts to punctuate the camera's inhibited vision at night. When shooting, I couldn't see my subject and they couldn't see me, so our other senses were heightened, and insightful conversations unfolded between us.

H: What has it been like to research in an IDSP and with dark sky communities/practitioners?

N: At the time of starting my research in 2016, the IDSP model was still very young. The prospect of researching from the situated perspective of a working IDSP and its stakeholders was both exciting and daunting. When I started fieldwork, I knew that the park did not have a dedicated team or working group but was surprised when I started conducting interviews to find that key stakeholders such as environmental agencies and conservation workers were struggling to connect with the GFDSP as both concept and material landscape. There was a sense that momentum had been lost and uncertainty around how people could get involved in the park's development. This was disheartening to encounter after the excitement of the EDSP conference and *Sanctuary* (Sanctuary, 2021) festival. The dark sky park, both as a concept and material landscape, felt nebulous, something that would prove difficult to "study".

As I got more involved with the GFDSP's stakeholders and landscapes however, its "nebulousness" became less of a concern and more a conceptual and methodological resource. If not in an official capacity, what other ways were people engaging with the dark sky park and how might these activities nurture stakeholder relationships and shape the future of the park? I joined stakeholders, practitioners, and residents on walks and site visits and spoke with them about their personal encounters and associations, often documenting their observations and experiences through sketches (Figure 2.3). Listening to a resident describe the pleasure of walking home from the pub with friends and their conversation being interrupted by a meteor shower attuned me to

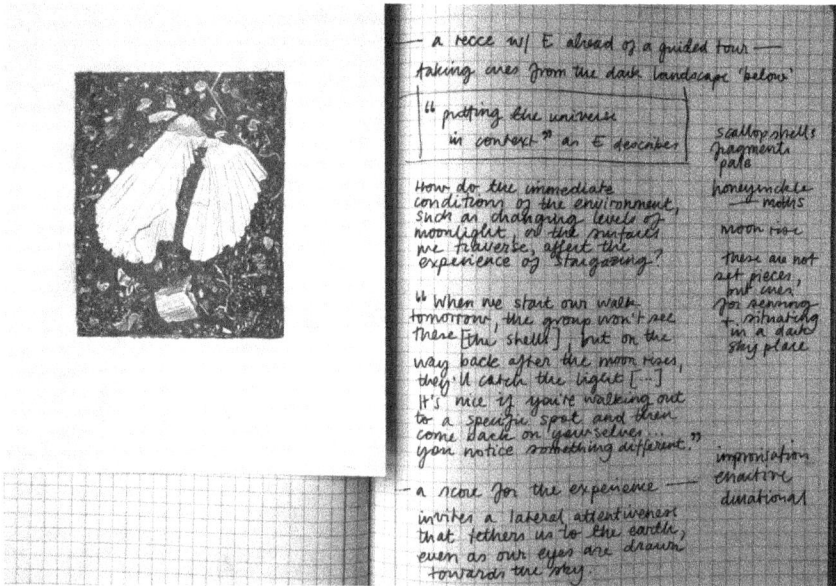

Figure 2.3 Illustration and notes reflecting on-site visit conducted with dark sky ranger ahead of hosting a guided tour. The illustration shows a smashed scallop shell used as surfacing material by Forest Enterprise Scotland, which then becomes an unusual texture in darkness and reflective detail in moonlight. Image credit: Natalie Marr.

the "everynight" experience of the GFDSP. On a site walk with a dark sky ranger to plan an experiential stargazing tour, I learned about how the GFDSP's practitioners incorporate Galloway's dark landscape (as well as its skies) in the visitor experience. On multiple site visits to a neolithic cairn across a 24-hour period, a devoted resident (and unofficial caretaker) invited me to explore the site through a series of sensory and conceptual exercises he uses as part of his daily visits (Figure 2.4). Such activities enabled me to map how dark sky values and "stakes" not only belong to conservationists and starry-eyed tourists but also emerge through everynight encounters and informal practices of dark sky engagement.

H: Working in Northumberland's immersive landscape was transformative, helping me to redefine my creative exploration of photography with the dark sky community. However, there were also moments that provoked both my creative process and KOAS operations, through the sometimes "disruptive" nature of art practices in rural places (Rowe, 2019). For instance, in February 2022, I installed an experimental "sonified image" (photographic data converted to sound) audio trail outdoors, which altered the "feeling" of the place (some visitors

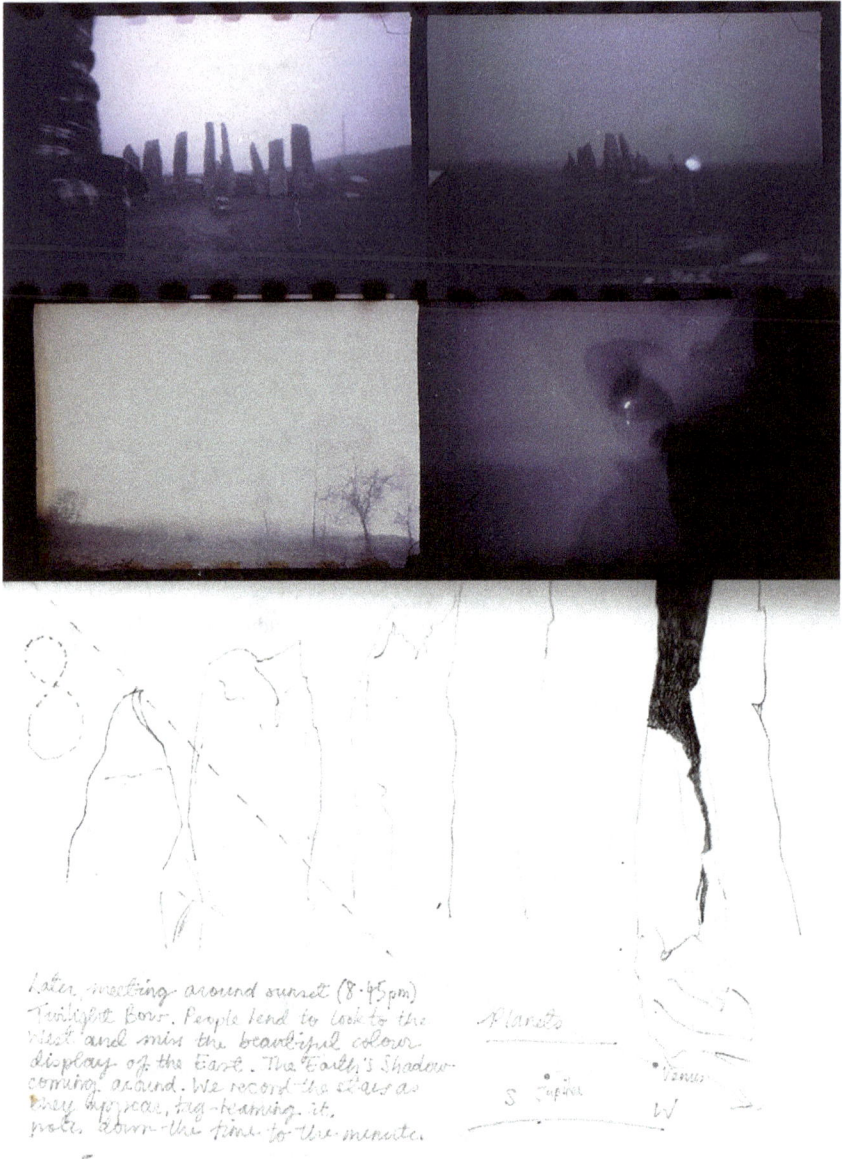

Figure 2.4 35 mm negatives made with a hand-constructed pinhole camera and sketches from visits to Cairn Holy with a local resident (2018). Image credit: Natalie Marr.

said it felt "eerie"), and my conversations with guests altered the event flow. The sound was my attempt to share photography without artificial illumination (regularly used in night-time artworks), so audiences could *imagine* pictures aided by the sound, rather than seeing them.

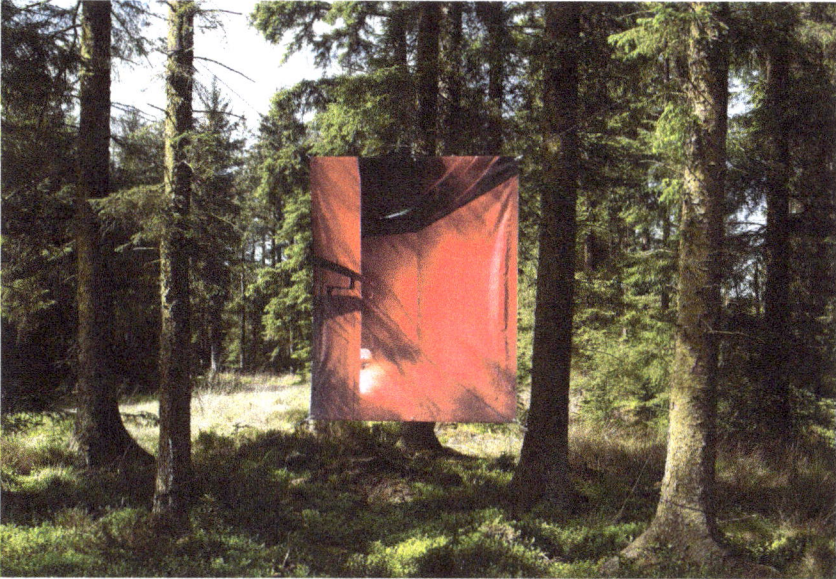

Figure 2.5 Installation of image "Approach" from *Another Dimension* outdoor exhibition at Kielder Forest, summer 2022. Eleven photographs were installed on the Skyspace Walking Trail towards KOAS. Image credit: Helen McGhie.

One visitor said she felt like an *explorer*, seeking to know what the sound was and said it intensified her experience; others were curious about the process, or recalled science fiction films. There was disruption to my practice too, which became solely focused on making work for KOAS rather than for "art world" audiences (as I had previously), in galleries and art publications. My research-led creative outcomes did not follow trends and were less interesting to art curators.

Seeking to engage Kielder visitors outside of stargazing season, I produced *Another Dimension*—an outdoor photography exhibition installed on the path to KOAS in summer (Kielder Observatory, 2022)—it offered a glimpse into a winter night, for those who happened upon it (Figures 2.5 and 2.6). One image, *Dark Adaptation* (Figure 2.2), pictured KOAS's car park, illuminated with the red "safelight" astronomers use to retain nocturnal sight, which may be likened to a photographic darkroom. I organised art walks, with KOAS trustees and volunteers, and co-hosted a public event with Peter Sharpe, who discussed the observatory's place in Kielder's Art and Architecture Trail (Figure 2.7). Attendees openly responded to the work; one visitor said that it "took them away from society and brought them serenity and peacefulness", others said the images were "otherworldly", and the exhibition made them "think about how each individual has their

Figure 2.6   Installation of image "Wanderer #2, 19.01.2019" from *Another Dimension* outdoor exhibition at Kielder Forest, summer 2022. Eleven photographs were installed on the Skyspace Walking Trail towards KOAS. Image credit: Helen McGhie.

own experience". Such mediations demonstrated how the artworks conveyed similar thoughts to those that KOAS staff had shared during my earlier photography shoots. I learned that audiences could feel something of the "dark sky" experience out of season, or on nights with poor visibility.

H: When working in the GFDSP, did you experience disruptions as an artist-researcher? How did this affect your creative practice?

N: Yes, disruption has been a valuable part of my practice, though not always in the ways I expected. I started fieldwork with a notion of the artist-researcher as "a disruptive force" who, by using alternative methods and modes of engagement, can creatively interrogate and expand "normative practices" (Hawkins, 2015, p. 156; see also, Edwards, Collins and Goto, 2016). There is certainly value in this, but it can also reinforce an imagination of research sites as stable and fixed locations, awaiting the researcher's interventions (Hyndman, 2001; Rose, 1997). It was as I was walking to or driving between locations at night that I experienced disruptions from the park's various nocturnal inhabitants in a literal way—flashes of wings and tails across my windscreen, the intrusive spotlights of "lampers" hunting deer and hares, and the

Figure 2.7 Documentation of the *Art Trail Walking Tour* led by Helen McGhie and Peter Sharpe, Kielder Forest, September 2022. Image credit: David Partington.

subtle but ever-changing variations of light and darkness in the forest as I made long-exposure photographs.

Such encounters opened new spaces of enquiry, shifting what was meaningful in the project and who or what participated in this meaning-making. This helped me to access a different experience of myself

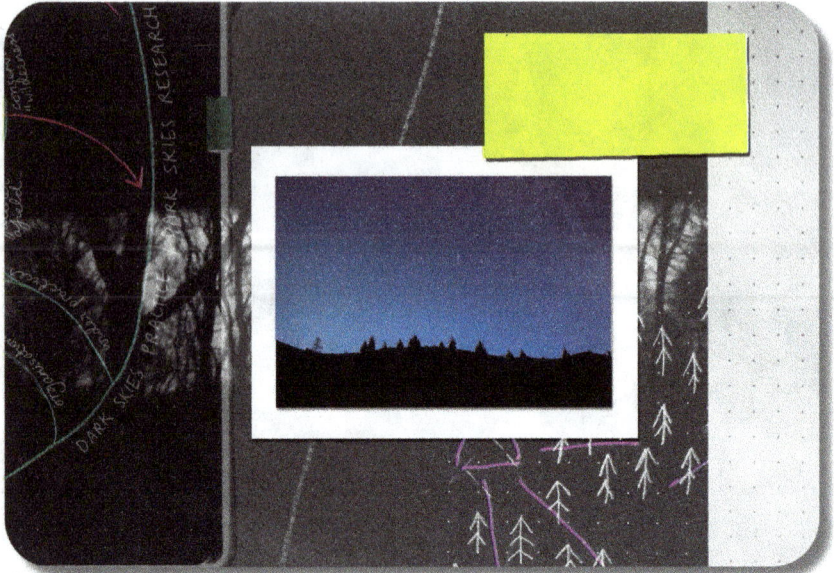

Stakeholders,

What stake do you have in research about dark skies?

What would you like researchers to investigate?

What would be helpful to understand about other dark sky places and practitioners?

How do your experience, skills and methods support or perhaps challenge dark sky research?

Researchers,

How might the GFDSP and other IDSPs be productive of international dark sky knowledge and practice?

How will you address the site? With what frames of reference? What modes of engagement?

What stake do you have in dark sky places and practice?

How can we work together or alongside?
What might a partnership look like?

Figure 2.8 A "prompt" card (front and back) included in an alternative field guide to the GFDSP, made in response to a growing sense of the researcher as a stakeholder, co-involved in the development of the park. Image credit: Natalie Marr.

within the unfolding narrative of the GFDSP's stewardship, by emphasising my capacity as an artist-researcher to be open and responsive to the generative potential of the site (Kester, 2004, p. 24). I explored this further through an evening workshop I facilitated with stakeholders in a wood located in the core zone of the park. The workshop was staged as "an alternative committee meeting", in which we explored Galloway's dark skies and landscape through a series of situated and sensory exercises that drew on my research encounters and creative approaches with an emphasis on receptivity and relationality as modes of stakeholder engagement. This included a "call and response" exercise using head torches to interact with one another from different positions in the wood and deep listening exercises (Figure 2.8). Participants also contributed to the workshop "programme": a staff member of Forestry and Land Scotland gave an impromptu talk about a site-specific artwork, and a dark sky ranger led a campfire activity at the end of the night. While the workshop invited participants to engage in unexpected ways with the GFDSP and with each other as dark sky stakeholders, it also opened up my own research engagements, especially my creative practice, which was normally planned and conducted alone. Sharing it in this live group setting encouraged me to think of my research as a form of stakeholding and of myself as a person with stakes, co-involved in the stewardship and development of the GFDSP [Figure 2.8].

N: As you are nearing the end of your research partnership with KOAS, does it feel useful to think of yourself as holding stakes within dark sky practice and community?

H: Yes, definitely, I am now part of the KOAS community, and in turn, they (and the environment) have shaped my approach to creative practice. My work occupies a "betweenness", crossing the worlds of photography and an IDSP, both of which have been enriched and challenged. I can relate to an astronomer's practice of "averted vision" as a useful metaphor for looking through peripheral sight to see something less luminous; I now look for less apparent contexts (and audiences) to creatively engage with and have invited KOAS to do the same, who have enhanced their public offer through the arts. My relationship with photography has shifted, and my motivations for making work feel more meaningful, especially since they engage with broader public audiences.

I am at the end of this long-term project and can see the contributions I have made to KOAS: expanding the events programme, facilitating conversations that allow reflection on organisational practice, and challenging what photography can be in an IDSP context. There were expectations of "what a KOAS photographer does"; they capture the night's sky. My creative practice —which reflects on the experiential *through photography,* rather than capturing the stars above—recalls KOAS's original transgressiveness within the landscape, it has shaped how visitors encounter the organisation. Engaging in dialogue with the KOAS community became essential for finding synergies and insights; the spoken *exchanges* between us were central and informed the work at each stage, it is ongoing. As a transformed practitioner, I now seek to apply this practice to new places, to make new connections and gather new knowledges.

Reflection

Through a reflective dialogue, we have shared how artist-researchers can work in partnership with IDSPs and the mutual benefits that may occur. Helen invited KOAS communities into her creative process through ongoing dialogue, from conversations with the community to artistic planning, dissemination, and reflection on her creative outcomes. Natalie engaged GFDSP stakeholders in collective reflection to map its diverse values, lifeworlds and practices in the years following international designation. Partnership work can disrupt the conventional practices of IDSPs and artists, leading to constructive outcomes. We now close the chapter with a set of critical reflections and propositions for future work.

There is value for IDSPs when working with artist-researchers. The creative process can enhance how existing communities (and new audiences) encounter dark sky places through new imaginative, emotional and sensory experiences that promote "public enjoyment of the night sky and its heritage" (International Dark-Sky Association, 2018 p. 3), whilst enriching narratives around the cultural values of dark skies. Further, our work contributes to the development of international dark sky places by engaging stakeholders in reflection on professional practice and their motivations for protecting the night. We encourage IDSPs to seek partnerships with artist-researchers and to allocate time and resources to build on our work.

There are benefits for artist-researchers who situate themselves in IDSPs. We have been ambitious when co-producing work with dark sky communities, by applying our creative skills to a new environment, beyond the "art world", whilst also responding to recent calls for interdisciplinary approaches in dark sky research (Kyba *et al.*, 2020). As creative practitioners whose work thrives on inspiration and (quite often) disruption, we have experienced IDSPs as generative contexts through which to develop work. We have experienced transformation in our practices, shifting from "artists" to "artist-researchers" with creative stakes in IDSPs which we frame not just as research contexts or case studies but sites of knowledge production themselves. We encourage future artist-researchers to work on-site to develop new interventions that respond to the intangible matter of darkness. Practice-based PhDs, in particular, may offer excellent opportunities for artists to develop work in and with dark sky places through sustained partnerships.

Our reflective dialogue affirms the value of sharing practice as artist-researchers developing work in a unique research context. We have reflected and elaborated on one another's independent research, demonstrating to ourselves and others the value of arts-based approaches to dark sky contexts. Through dialogue, we have shown that whilst we started with a similar collaborative setup, our methods and outcomes differed as our explorative work responded to the shifting IDSP seasons, engaged in conversations with stakeholders, and informed public outreach activity. We propose further opportunities for dialogue between artist-researchers, through conferences or workshops that can provide space for those who have worked in dark sky contexts to share and reflect on their experiences, whilst introducing this rich area to those wishing to develop projects with dark sky places. Such conversations would build capacity among artist-researchers, strengthen the case for further creative interventions in IDSPs, and inform dark sky stakeholders on the resources required.

References

Adams, A. (1948 (2002)) *The negative*. Reprint. London: Little, Brown and Company.

Arts at CERN. (2023) *Artistic residencies*. Available at: https://arts.cern/programme/artistic-residencies (Accessed: 5 January 2023).

An Lanntair. (2022) *Hebridean dark skies festival*. Available at: https://lanntair.com/creative-programme/darkskies/ (Accessed: 12 January 2023).

Barrett, E. and Bolt, B. (eds.) (2010) *Practice as research: Approaches to creative arts enquiry*. London: I.B. Tauris.

Blair, A. (2016) *Sark in the dark. Wellbeing and community on the dark sky island of sark*. Lampeter: Sophia Centre Press.

Candy, L. (2019) *The creative reflective practitioner: Research through making and practice*. London: Routledge.

Charlier, B. (2018) '"You know the pyrenees by day—come see them by night . . ." Reflections on in visu Artialisation of nocturnal skyscapes in the Pyrenees', *Journal of Alpine Research*, 106(1), pp. 1–24.

Cotton, C. (2020) *The photograph as contemporary art*. London: Thames and Hudson.

Edwards, D. M., Collins, T. M. and Goto, R. (2016) 'An arts-led dialogue to elicit shared, plural and cultural values of ecosystems', *Ecosystem Services*, 21(November), pp. 319–328.

European Space Agency. (2023) *Artist in residence*. Available at: www.cosmos.esa.int/web/artist-in-residence/ (Accessed: 5 January 2023).

Finlay, A. (2008a) *Time bends space arcs light eclipses* [windmill turbine with poem]. Kielder Observatory. (Viewed: 2017).

Finlay, A. (2008b) *One hundred year star-diary*. Chorley: Platform Projects.

Glinkowski, P. and Bamford, A. (2009) *Insight and exchange: An evaluation of the wellcome trust's Sciart programme*. Available at: www.wellcome.ac.uk/sciartevaluation (Accessed: 4 January 2023).

Hawkins, H. (2015) *For creative geographies: Geography, visual arts and the making of worlds*. London: Routledge.

Hyndman, J. (2001) 'The field as here and now, not there and then', *Geographical Review*, 91(1–2), pp. 262–272.

International Dark-Sky Association. (2018) *International dark sky park program guidelines*. Available at: www.darksky.org/wp-content/uploads/bsk-pdf-manager/2021/05/IDSP-Final-May-2021.pdf (Accessed: 2 January 2023).

Jafari, J. (2007) 'Terrestrial outreach: Living the stardome on earth', in Jafari, J. and Marín, C. (eds.) *Starlight: A common heritage*. Tenerife: Astrophysical Institute of the Canary Islands, pp. 55–57.

Jerram, L. (n.d.) *Museum of the moon*. Available at: https://my-moon.org/ (Accessed: 31 January 2023).

Kester, G. H. (2004) *Conversation pieces. Community and communication in modern art*. London: University of California Press.

Kielder Art and Architecture. (2023) *Home*. Available at: http://kielderartandarchitecture. visitkielder.com/home.html (Accessed: 23 January 2023).

Kielder Observatory. (2021) *History*. Available at: https://kielderobservatory.org/the-obser vatory/our-history (Accessed: 30 February 2023).

Kielder Observatory. (2022) *Another dimension outdoor photography exhibit*. Available at: https://kielderobservatory.org/news/latest-news/235-another-dimension-outdoor-photog raphy-exhibit (Accessed: 10 October 2022).

King, M. (2022) *Ancient light: Silver gelatin photographs, 2017–2022*. Available at: www. melaniek.co.uk/ancient-light-rematerialising-the-astronomical-image#! (Accessed: 1 December 2022).

Kyba, C. C. M., Pritchard, S. B., Ekirch, A. R., Eldridge, A., Jechow, A., Preiser, C., Kunz, D., Henckel, D., Hölker, F., Barentine, J., Berge, J, Meier, J., Gwiazdzinski, L., Spitschan, M., Milan, M., Bach, S., Schroer, S. and Straw, W. (2020) 'Night matters—why the interdisciplinary field of "night studies" is needed', *Journal: Multidisciplinary Scientific Journal*, 3(1), pp. 1–6.

Mayo Dark Sky Festival. (n.d.) *About the mayo dark sky festival*. Available at: www.mayo-darkskyfestival.ie/about (Accessed: 12 November 2022).

McGhie, H. (2020) Conversation with Peter Sharpe, 4 May.

National Park Service. (2022) *Imma Barrera*. Available at: www.nps.gov/acad/getinvolved/ air-barrera.htm (Accessed: 5 January 2023).

Rickett, S. (2014) *Objects in the field*. Available at: https://sophyrickett.com/objects-in-the-field-1 (Accessed: 2 October 2022).

Rose, G. (1997) 'Situating knowledges: Positionality, reflexivities and other tactics', *Progress in Human Geography*, 21(3), pp. 305–320.

Rowe, F. (2019) *Experiencing contemporary arts organizations in rural places: Arts practices and disruption in North East England and Scotland*. PhD thesis. Newcastle University. Available at: https://theses.ncl.ac.uk/jspui/handle/10443/4736 (Accessed: 13 December 2022).

Sanctuary. (2021) *About*. Available at: https://sanctuarylab.org/about/ (Accessed: 5 January 2023).

Sumartojo, S. and Pink, S. (2019) *Atmospheres and the experiential world. Theory and methods*. London: Routledge.

Vannini, P. (2015) *Non-representational methodologies. Re-envisioning research*. London: Routledge.

Visit Kielder. (2000) *Cat cairn: The Kielder Skyspace—James Turrell 2000*. Available at: www.visitkielder.com/great-outdoors/cat-cairn-the-kielder-skyspace-james-turrell-2000 (Accessed: 5 January 2023).

3 Dark skies in southern Scotland and northern England

Border-crossing sites for creative experiment and envisioning connectedness

Ysanne Holt

Throughout centuries of feuding, the often-harsh terrain typical of the cross-border region of northern England and southern Scotland was frequently lawless, wild, and dangerous. For thieving Border Reivers of the late 13th to early 17th century, the pitch-black skies of the long winter months concealed night-time cattle rustling, blood-thirsty acts of vengeance and blackmail, and hasty retreats over mist-cloaked moorland and hills and through dense dark forest. The Scottish historian Leslie, Bishop of Ross, famously described the customary behaviours of inhabitants of the marches of both kingdoms in 1607:

> They sally out of their own Borders, in the night, through unfrequented by-ways and many intricate windings. All the day time, they refresh themselves and their horses in lurking holes they have pitched upon before, till they arrive in the dark at those places they have a design upon. As soon as they have seized upon the booty, they in like manner, return home in the night, through blind ways, and fetching many a compass. The more skilful any captain is to pass through those wild deserts, crooked turnings, and deep precipices in the thickest of mists and darkness his reputation is the greater.
>
> (Borland, 2010, p. 59)

There is much associated place-making here, as, for example, in the influence of Walter Scott's *Minstrelsy of the Scottish Border* (1802), in a lingering legacy of darkness related poetry, stories, and song marking violence, heroism, and romance, or unsettling, nerve-tingling legends of supernatural apparition and shape-shifting metamorphosis. Related forms of tangible and intangible heritage have ever-since been produced and perpetuated across the Scottish borders, north Northumberland, and Cumbria, contributing to what is understood here as the performance of a cross-border region, a borderland. This borderland, often historically regarded as socially, culturally, and economically peripheral and isolated, is a sparsely populated region with shared material resources, customs, skills, and traditions emerging from its extensive areas of uncultivated moors and mires, dispersed upland farms, large swathes of forestry plantation and, at its coastal edges, declining fishing, and

DOI: 10.4324/9781003408444-5
This chapter has been made available under a CC-BY-NC-ND 4.0 license

maritime communities. In the present day its dark skies above, however, are especially valued as pristine sites, preciously free of the insidious orange glow of modern-day pollution from artificial light sources emanating from urban areas, roads, and traffic. Such polluted conditions would have confounded the Reivers in their nightly pursuits and obscured their clear view of the star-filled sky. Recognised today as natural assets to the region, as much as its generally clean air, water, and soil, these characteristically dark skies contribute to an overall sense of tranquillity, scenic beauty, and healthy environment.

Considerable sections on either side of the border are now officially recognised as dark sky parks. In Scotland, the Galloway Forest, the largest forest in the UK at 774 square kilometres, was designated the International Dark-Sky Association's fourth dark sky park in the world and the first in the UK and awarded Gold Tier status in 2009. Across into England and the 572 square miles of Northumberland's Dark Sky Park is the largest in the UK and the second largest in the whole of Europe, spanning as it does the Northumberland National Park and the Kielder Water and Forest Park up at the northern border with Scotland. Both have become central to the region's tourist economy, creating new experiences, and bringing new audiences.

In Northumberland, a range of viewing sites include the county's five observatories; one of which is at Stonehaugh, a village built in 1957 to house Forestry Commission workers, its function now adapted to recreational amenity rather than work circumstances—so to stargazing, as well as walking and cycling trails. Another, the Battlesteads Dark Sky Observatory, near the village of Wark, is a facility for educational talks and astrophotography workshops. The Northumberland Dark Skies Festival in February 2022 saw activities across all five observatories (the remaining three are at Kielder, at the Twice Brewed Inn on Hadrian's Wall, and at Allenheads in the North Pennines) as well as the Northumberland National Park Authority's (NNPA) Dark Skies exhibition at the Sill, its National Landscape Discovery Centre, also on the route of Hadrian's Wall. The Sill featured a "gallery of night-time landscape photography, displays of what to see in the night sky and how to conserve our dark skies, with hands-on activities and a showcase of telescopes and meteorites on display" (Hadrians Wall Festival, 2022). All of this met the Centre's obligations to its initial funder, Heritage Lottery, to engage multiple and diverse audiences in the exploration and better understanding of landscape and environment; past, present, and future.

Since the early 1990s, Northumberland's Kielder Water and Forest Park has been a significant site for public art, initially of commissioned sculptures and later architectural installations in the immediate environs of the vast reservoir that was constructed in Forestry Commission land and opened in 1982 to provide water to industrial Tyneside, Wearside, and Teesside. Following increased patterns of partnership activity between local authorities, conservation agencies, creative practitioners and arts funding bodies, Kielder "dark skies narratives", as experienced and shared through storytelling amongst the communities who live and work in the surrounding area of the North Tyne, are currently being explored with the support of funded artist researchers (Smith, 2021). The arts here are seen to articulate place-making values, to facilitate expression of a local community's identity and sense of place, as well as enhancing the tourist economy and, as it is claimed in this instance, develop employment opportunities. More generally here, as elsewhere, the dark skies experience is also broadly perceived as a valuable corrective to modern "nature deficit disorder", to the promotion of personal and collective wellbeing benefits as well as wider environmental awareness. Artists' engagements, as in this case at Kielder, are frequently understood and expressly funded to contribute to the delivery of wide-ranging social, cultural, environmental, and health policy objectives.

In that context, and as returned to in the following, certain contested debates do arise over the implicit instrumentalisation of arts practice at the expense of what might be regarded as its proper aesthetic concerns. What follows engages with certain aspects of this debate. The focus overall, however, is the extent to which the dark skies parks of Galloway and of Northumberland have over the past decade or so emerged as innovative spaces for critically engaging diverse constituencies through experimental practice in forms of performance, interactive installation, sound works, and video projections. Through consideration of examples of these, the chapter considers the potentialities and implications of this borderland's dark skies as a way of thinking across and beyond the constraints of borders and boundaries and envisioning new forms of connectedness and being together. It takes the notion of border-crossing conceptually as well as geographically, viewing the crossing between disciplines, practices, and bodies of knowledge as a distinctly generative process. Its focus is upon creative entanglements with the immaterial, as well as the material resources of the cross-border region.

Those forms of creative engagement reflect a general shift in recent decades in the nature of public art, or art in public space, and its status and concerns. The tendency has moved away from structures often understood as "imposed" on sites towards more dematerialised or immaterial art forms and temporary experiences and interactions, witnessed in a trajectory from certain traditions of conceptual, land, or environmental art. As curator and art historian Miwon Kwon terms it, there has been a "shift from site-oriented practices towards the dematerialization of the site" to a "specific relationship between an artwork and 'its site' not based on the physical permanence of that relationship but a recognition of its unfixed impermanence to be experienced as an unrepeatable and fleeting situation" (Kwon, 1997, p. 91)

Following such a pattern, in this northern region there has been a marked transition from the type of sited artworks resulting from initial Forestry Commission partnerships and artist residencies from the 1970s at Grizedale Forest Sculpture Park in Cumbria and from the 1990s at Kielder. This includes those art works expressly intended to decay back into the natural environment and exist long term only through forms of documentation. Instead, more socially engaged practices have predominated with participation with publics and local communities frequently at their core, often as collaborators or co-producers. In some instances, these refer us back to the more overtly politicised community arts movement from the late 1960s. The emphasis in this context is commonly on process rather than product. Such an emphasis parallels the understanding here of landscape itself as always "in process", as experiential and relational rather than a fixed view or a backdrop. In terms of the latter and in tandem with non- or more-than-representational thought emerging from disciplines such as social anthropology and cultural and human geography, those shifts in much contemporary art practice away from painting and certain types of traditional landscape photography, for example, also meet with wider concern over an overt privileging of the visual at the expense of other sensory responses—touch, taste, sound, and smell—that support a fully embodied experience of being in, rather than simply looking at, landscape or environment. In this context the term "landscapism" or "landscape as milieu", as considered by cultural geographers Tim Edensor and Haydn Lorimer (2015, p. 1), aptly defines a shared focus in recent performance events as in discussions on the non-representational within the social sciences with what they term an "estranging encounter with topography, terrain and atmosphere". So to engage audiences as participants in diverse forms of performance or installation, as in the following examples, is potentially more capable of fostering an ecological sense of the fully entangled nature of human and non-human interactions within landscape.

Alongside these forms of dematerialised art, this understanding of landscape in the context of the apparent boundlessness and the indifference to boundaries of the dark skies is especially pertinent in terms of the framing concern here with the dismantling of borders and acts of bordering. Dark skies, like the circulation of air and the immensity of the sea, consistently confound the imposition of divisions and the attribution of singular defining identities. This is to follow Doreen Massey's influential articulation of a global sense of place. For Massey, geographical space is borne from "a constellation of social relations, meeting and weaving together at a particular locus" (Massey, 1994, p. 154). In the context of time-space compression, all places are perceived here as a mix of local and wider social relations and as outward, unique, and progressive rather than self-closing and defensive. In the terms of Kwon's commentary on the dematerialisation of site-based practices, Massey's view of place underlines the importance of cultural practices bearing a relational specificity, of recognising the constellation of social relations within a particular environment, rather than emphasising any one particular identity. In the context, too, of new materialism or the eco-philosophy of figures such as Karen Barad, those relations and identities extend of course beyond the merely human, to "intra-actions" with non-human actors (Barad, 2007, p. 149).

Above, below, and beyond

One of Northumberland's five community observatories, referenced earlier, has been located since 2017 at the village of Allenheads, at the site of Allenheads Contemporary Arts (ACA). ACA was established in an old schoolhouse in 1995 by artists and curators Helen Ratcliffe and Alan Smith as an "integration of life, work and contemporary art in a rural location", a "place to operate through dialogue", in an area "shaped by successive human generations and their manipulation of matter as well as by natural forces" (Ratcliffe and Smith, 2009, pp. 7–9). Across this time ACA, at times in partnership with the local AONB, has supported several art and science collaborations such as the programme *As Above So Below*, 2016. This encouraged local and international artists in the exploration of both the dark skies above and the subterranean environment below—the hollow earth—through experiential walks and periods of immersion in the disused lead mines around the village, often resulting, for example, in field-recordings supporting a collective listening, and to diverse collaborations between artists, writers, and musicians. ACA's residency programme was perceived as both "testing ground and an incubator" and an opportunity for artists to develop work without boundaries (Allenheads Contemporary Arts, 2018).

To that end, the Dumfries and Galloway based public artists Robbie Coleman and Jo Hodges were selected as residents in 2016. Coleman is an artist and curator with a focus on sculpture and live art as well as a manager and co-director of large-scale public arts projects, whereas Hodges, also an artist, curator, and producer, has a background in human ecology, community development, and social justice. Their practice is innately multi- and cross-disciplinary in its engagement, as they describe, with "ecological and socio-cultural systems, processes, relationships and change", their concerns "spanning environment, technology, culture, conflict, adaptation and resilience, speculative futures, global relationships to the local, participatory culture and the role of art in social and ecological change". Consequently, Coleman and Hodges' research, experimentation, and collaboration has taken numerous forms, resulting in "temporary and permanent works, site specific installations, socially engaged and participatory processes and explorations of new strategies for working in public space" (Coleman and Hodges, 2023).

The culmination of these artists' ACA residency is especially pertinent in the comprehensions here of landscape as processual, on the dematerialisation of site, on place as a constellation of diverse human and non-human actors and identities, and emphases on multi-sensory and embodied experience through active participation rather than passive spectatorship. This was TETHER (Tenuous Links, Universal Truths and Far Away Things: A sensory journey into the dark to explore the forces at play in unseen and in-between spaces). TETHER was a participatory, experiential walk that took place below the dark skies of the North Pennines (Figure 3.1). As such, it asserted the importance of a sensuous and slow unfolding of the environment. An established tradition within contexts of both land art and performance, more generally, walking has been valued as a "way of becoming unstuck from yourself, of merging with your environment; the boundary between yourself and the environment is relinquished . . . It is about a dematerialisation of the self, dissolution into space, becoming part of the geography" (Van Den Boogaard, 2014, p. 120). In that regard walking was key to Coleman and Hodges' exploration, along with their participants, of the human relationship to "darkness, place, space and each other" (Coleman and Hodges, 2023).

As the artists documented, throughout their journey, those engaged in TETHER encountered performative elements and experiences that underlined the notion of darkness as both place and concept in the present and across time. The sense of crossing thresholds was witnessed through references to the rowan tree, the ancient

Figure 3.1   TETHER, Jo Hodges, and Robbie Coleman at *As Above So Below*, Allenheads
Contemporary Arts, 2016. Photo by Nat Wilkins.

wayfarers' tree and a pathway for the movement of materials—a physical bridge
between the surface and underground worlds. Sensory awareness of that movement
and connection was heightened through the group's tasting of rowan jelly, here
made from berries gathered from a rowan tree at Allenheads, mixed with apples
from a tree at the artists' home across the border in Scotland. Linkages between
this place and others across the world, out in space, back in time, and over diverse
cultures were evidenced as the participants walked out in darkness, in silence and
in single file from the village of Allenheads towards nearby woodland. Before
entering the wood, they were introduced to an event that happened in a different
time and place. As they walked into the woods, they listened to the soundtrack of
American blues singer Blind Willie Johnston's "Dark Was the Night, Cold Was the
Ground"; they walked further on to a clearing where an installed screen showed
the launch of the deep space probe Voyager 1 of 40 years earlier. At that point, in
2016, Voyager was 12 billion miles from home beyond the edge of our solar system. It was the furthest human made object from the earth and still sending information back. A golden record containing information from planet earth, including
Johnston's song, about loneliness, had been bolted to the side of the probe before
launch. Each participant moved to a place in the woodland where they stood alone
for a while in silence in the darkness. They then walked slowly in single file in the
dark back into the village. On their return the group tasted homemade Cantonese
mooncakes—a feature of the Chinese celebration of the mid-autumn full moon.
These mooncakes had a star imprint, which was the signature pattern of the early

19th-century local Allenheads quiltmaker, Elizabeth Sanderson. Coleman and Hodges had taught themselves the local quilting technique, incorporating Sanderson's star and the code from the Voyager 1 record needed to extract the information it contains—sounds, images, and greetings. TETHER participants were invited to place stitches into the quilt, a tactile collective contribution. Echoing Massey to some degree, too, in considering the significance of their *As Above* residency, the artists quoted Tim Ingold: "To be, is not to be in place but to be along paths. The path, and not the place, is the primary condition of being, or rather of becoming" (Ingold, 2011, p. 12).

In the context of planning and choreographing their night-time event and sharing in the experience with others, Coleman and Hodges were underlining the distinct value of performance as a means of attending to human and non-human interrelations and mobilities, both material and immaterial and linked across time, place, and space. Glasgow-based theatre and performance artist Minty Donald has described her own and Nick Millar's water-related scores and events as an "ecological practice". In similar terms here, TETHER challenged representationalism, blurred distinctions between "everyday" and "art" performance, between performance and performativity, manifesting in the "sensorial, affective and more than cognitive" (2016, p. 253).

As integral to TETHER, sound or sonic practices in general and the construction of aural topographies frequently feature within creative responses to the dark skies and speculation on the beyond. (*Beyond* was the title of another ACA programme, this one in 2018, just after the opening of the observatory, and to which Coleman and Hodges also contributed.) In the context of sound, Newcastle University–based artist-researchers Tim Shaw and John Bowers have produced "A Midsummer Night's Drone" at ACA, a simple harmonic structure accompanying the longest day from dawn to dusk and the cycle of the sun around the earth. Mindful of the diversity of sonic practices and returning to that (at times possibly over-stated) critique of an excessive focus on the visual in notions of landscape, the environmental sound artist Bennett Hogg, another frequent contributor to ACA events, has argued for sound as offering a phenomenological alternative to visual dominance. Hogg refers to Salome Voegelin with the claim that the sonic "opens up more explicit possibilities for sensory integration than the visual, the aural sense being more closely imbricated with touch, vision, movement, and muscular reaction" (Hogg, 2018, p. 239; Voegelin, 2010).

Sanctuary and the dark outside

Ecological or environmental thinking has been consistently apparent in creative engagements with dark skies across the border from Allenheads to southern Scotland, and sonic forms have featured amongst these. Coleman and Hodges played a significant part in the establishment of the Environmental Art Festival Scotland. This evolved from 2012 with Dumfries and Galloway arts organisations connecting landscape, environmental and countryside agencies along with networks of local and international art practitioners. Two resulting festivals in 2013 and 2015 included numerous multidisciplinary environmental art projects, events, immersive experiences, and installations, using the "local context to explore global issues of environmental consciousness and land and to consider future ways of living and thinking for those who live and work in the region" (EAF Scotland, 2015).

Emerging from this context was *Sanctuary Lab*, a 24-hour public art laboratory and experimental space in the GFDSP with Coleman and Hodges as Director-Curators. *Sanctuary* (as it's described, from both light pollution and surveillance) was to bring together artists and publics interested in place, light, sound, technology, landscape, and environment "to experience experimental artworks, digital and sound works, radio transmissions, video, talks, workshops and performance" (Coleman and Hodges, 2021). The challenge, or rather the opportunity, presented by *Sanctuary* as it manifested over ten years was experimentation with what might be possible in a dark, hidden space at an edge, with being in a not networked place, a place within which to explore new technologies or new ways people might "be" with technologies—an "off-grid" site, far from a city, with diverse groups of people coming together and sharing in affective experience. Little was recorded or documented, and there was no interest in the creation of spectacle.

Related to the development of *Sanctuary* was *The Dark Outside*, an event in the GFDSP first initiated in 2012. At that event and for several years later, Coleman and Hodges collaborated with sound artist Stuart McLean who developed the 24-hour site-specific radio broadcast of never-before-heard "largely drone ambient" soundtracks gathered from international musicians. This was broadcast from Talnotry at the site of the monument to the late-18th- and early-19th-century linguist, scholar, and son of a local shepherd, Alexander Murray. The site, McLean remembered, was chosen specifically as "FM works from line-of-sight, which hems the listener (due to the geology and landscape) into a specific area and requires audiences to travel out into a forest to listen to the radio" (McClean, 2017). An overall concern was to avoid what he termed in a review the "corporate-sponsored festival route", observing that the event would be ruined if it "became something which needed security teams and people in yellow jackets directing cars around . . . maybe we just need a few more inflatable chairs" (Turner, 2013).

In 2015, *Sanctuary* took place from noon to noon on 25 and 26 September, again at Talnotry. The notion of darkness here extended to include electronic darkness: "Sanctuary provided a temporary escape into digital darkness where our relationship with all kinds of technology was explored in experimental and creative ways" (Coleman and Hodges, 2021). Crucially here, and as stated, *Sanctuary* explicitly did not reject the digital age. There was no implication of retreat into an anti-urban rural world associated with the past. Those artists and participants involved created digital artworks and networks on the actual site, in a context which encouraged critical reflection on the "meaning, uses and implications of technology and the ownership and agency of the devices and networks that now connect us" (Coleman and Hodges, 2021). Such a concern seems to resonate with the post-phenomenological thought associated most influentially with Don Ihde, in which far from a rather romantic stance of distance or even dismissal of technology, technologies are analysed as significant mediators of human experiences and practices (Ihde, 1993).

The nature of the digital and the live event here give rise to further consideration of notions of site or of place. Valuable observation in this context emerged from a 2016 published conversation between filmmaker and anthropologist Julia Yezbick and the sound artist Mark Peter Wright. Wright questions what "site" might mean for an individual like himself who spends time in a laptop, thus "what does it mean to treat the screen as an environment", or for Yezbik, "what does site specificity mean when we have the ability to extract images and or/sounds and circulate them within these new networks of exchange and mediation". Wright underlines what appear key concerns for *Sanctuary* and with echoes of both Kwon and Massey. His problem is with any essentialist desire to "represent site" as in the earlier context of Land Art. His focus is rather to "propel site specificity into something more inclusive and accessible than the purely phenomenological 'being there' approach". Wright is influenced here by Mary Louise Pratt's thinking in which the field is perceived as a contact zone. Here

> the physical site is transformed into a participatory arena where human and non-human relations are performed and enacted. The field becomes a contact zone . . . as a result site becomes more about how an environment, animals and technology capture and "do" upon me.
>
> (Yezbik, 2016)

There is synergy here too with the practice of Tim Shaw, mentioned earlier, who was commissioned by Coleman and Hodges for *Sanctuary* in both 2015 and 2017. Throughout Shaw's engagement with performance installation, DIY electronic music, improvisation, and maker culture are constructions and investigations of affective environments (Shaw, 2023). As such his 2015 *Sanctuary* commission allowed for a version of *Fields* (the title a reference to field as site, as arena, but also the method of field recording) an ongoing series of sound performances produced in collaboration with experimental musician and programmer, Sebastien Piquemal. In this context, *Murmurate* (the title of their *Sanctuary* event), sound was relayed via a set up Wi-Fi network through the mobile devices of event participants (Figure 3.2). The *Fields* series was responsive to each specific performance context. In this instance the off-grid, digitally dark, and quiet environment of the Galloway Forest was appropriate for an event based on the use of mobile phones. It was an opportunity to develop a one-off, collective, collaborative experience, marked by a strong element of unpredictability. Shaw himself has underlined connections with Ingold here in terms of an interest in "wayfaring", in improvising with materials found along the way (Richards and Shaw, 2022). Interestingly, however, and connecting with the reference to post phenomenology, there is a degree of challenge in Shaw's assertion of the digital as material, in contrast to what in Ingold can appear to be an overly romantic emphasis on material in terms of stone, wood, and so on, and on "making" in the context of traditional craft, rather than with contemporary technology.

The location for *Murmurate* had no power at hand, so a system had to be run from two car batteries and a leisure inverter carried to it, somewhat ironically

Figure 3.2 Murmurate, Tim Shaw, and Sébastien Piquemal at Sanctuary Festival, 2015.
        Photo by Alison Boyes.

referencing manual labour in interaction with materials. Field recordings were collected from the site in advance of the performance, and these included sound from a nearby stream gathered with the use of a hydrophone. As Shaw records, in this specific method which they termed "streaming a stream", subtle changes in the water's ebb and flow modulated a synthesised voice during the live performance. How the resulting sound was experienced by those who made their way to the event varied from individuals to self-selected groups forming and interacting around the site. Here the method of field recording and the engagement with participants underlines Ingoldian emphasis on the relationship between materials (here technology) and maker as ongoing, fluid, and responsive and as manifesting the entangled relations between the human and non-human. As Shaw notes too, Ingold emphasises the importance in the context of making of inevitable irregularities and imperfection. In the digital field, there is a parallel in the potential delay or latency in sound distribution. In relation to the series *Fields*, the instances and effects of networked latency are valued as creative material, unique to each individual space, to the local network, and to the overall experience, and not to be eliminated in any pursuit of seamless transition (Shaw, Piquemal and Bowers, 2015, p. 282).

In 2017 Shaw developed work for Reach, a project conceived by Coleman and Hodges with funding from the Space (an innovative and experimental Arts Council and BBC digital arts media service programme). The commission was to build up a bank of responses from a public call out to be uploaded onto a collaborative platform before a 24-hour event, *Going Dark*, to take place over 23–24 September. The final artwork, drawing upon the contributions, was therefore entirely the result of online participation and the aim overall was to capture public responses to the key themes. Shaw had made clear that he was interested in experimenting with communication and connectivity in order to explore darkness and the idea that a break with technology is a way of "going dark". The call asked people to "document darkness, record remoteness, and to send silence", and the collected sounds and images were transmitted on screens between two short range radio masts. As the environmental conditions on-site—temperature, rainfall, humidity, and so on—changed throughout the event, so the images were transformed and distorted by natural interference. In this way the material and immaterial conditions of the specific site in the dark sky park intermingle with multiple responses to the theme as contributed by individuals either close by or, potentially, far away who may well not have attended the final 24-hour event but had nonetheless actively contributed to its form and character.

In a wider context, in a review titled "Radio Activity", the artist and writer Lauren Velvick identified several exhibitions and events that had taken place in the winter of 2015 using "broadcast and performance to form temporary sites of participation and investigate the possibilities of sound to be both transitory and archivable". Velvick cited Grace Schwindt's "Little Birds and a Demon" live-streamed from an isolated lighthouse on the Shetland Islands to six venues in northern England and Scotland; curator Marion Harrison and artist Sophie Mallet's "Project Radio" in Leeds and Ross Jardine and Matthew de Kersaint Giraudieu's "The Map is the Territory" were delivered in Sheffield in December that year. For the latter, the use of an FM signal was integral to the production of site and community; together the artists spoke of mapping an area through an FM signal but also of simultaneously broadcasting online and reaching a much larger audience. At this same event, a playful, improvisational live-action role-playing (LARP), "actively demonstrated the potential power of an audience's collective imagination to conjure a new place into being" (Velvick, 2016, p. 32).

These demonstrations of the potential of forms of participatory practice in generating or facilitating critical reflection and intervention, resulting in a collective sense of possible alternatives to present circumstances, go some way to counter art historian and critic Claire Bishop's concerns that participatory, or socially engaged art practices can often amount to an over preoccupation with ethics rather than with aesthetics. Bishop identifies a tendency to focus on good or bad models of collaboration as the primary criteria for judging the success of an event or a particular project rather than the "disruptive specificity of a given practice" in terms of the "what makes it art element; the achievement of making social dialogue a medium; the significance of dematerialising a work of art into a social process, or the specific affective intensity of social exchange" (Bishop, 2012, pp. 22–23). Relatedly here, Thomas McMullan, in his Alphr.com discussion with Shaw about radio-television identified the current wider social and cultural interest at that point in time in "going dark", in awareness of the negative effects of 24-hour social media and the appreciation (or potential fetishisation) of disconnection (McMullen, 2017). However, in the events considered here, motivated participants engage as co-producers in a live arena for action, in improvisational, unpredictable contexts. They are therefore able to influence resulting artworks, to shape a collective experience building on available resources on-site or with materials (sounds, images, reflections) from elsewhere, and so to form affective relations and emotional connection within isolated contexts of apparent disconnection, far from centres of power or authority through engagement with now commonplace, everyday forms of mass-produced technology that are, as stated, more usually associated either with control or surveillance, or simply as the means to solve problems.

Salome Voegelin underlines what are identified throughout this chapter as the possibilities of soundscape and sonic *works* to help construct sonic *worlds*, or immersive environments as in phenomenological life-worlds, worlds that challenge or augment conceptions of the "actual world" and suggest alternative ways to be (Voegelin, 2013). As the diverse practices of Shaw, Hodges, and Coleman demonstrate, in a larger context of ethics and agency such a scenario requires continual questioning of the relationship between people and digital technology and an understanding of the need for more open systems. Relational concepts of landscape, place, and site, alongside diverse and collaborative activities emphasising an ecological sense of the interrelations of human and non-human actors and material and immaterial forces, are vital to these endeavours.

In this context, those relatively still isolated, sparsely populated regions of the dark sky parks of the north of England and southern Scotland, those "out of reach" areas with their unique resources have proved to be an ideal deliberative space, a cross-border borderland, an "arena" in which to investigate, in an unregulated manner, our complex relationships with technology and to engage in a process of unbounded "de-bordering", imagining future alternatives, better forms of connectivity and a greater embodied awareness of our own existence within fundamentally fragile and vulnerable environments. It is a very different and a future-oriented place-making, one in marked contrast to those historic associations with the region's dark skies of the Border Reivers. All this ultimately speaks to the continual need to demonstrate the wider significance and implications of interconnected ways of thinking, of being and doing within and across sites, fields, borders, and boundaries.

References

Allenheads Contemporary Arts. (2018) *As above so below*. Available at: www.acart.org.uk/as-above-so-below.

Barad, K. (2007) *Meeting the universe halfway: Quantam physics and the entanglement of matter and meaning*. Durham: Duke University Press.

Bishop, C. (2012) *Artificial hells, participatory art and the politics of spectatorship*. London: Verso.

Borland, R. (2010) *The project Gutenberg E-book of border raids and Reivers*. Available at: www.gutenberg.org/files/32005/32005-h/32005-h.htm.

Coleman, R. and Hodges, J. (2021) Available at: https://sanctuarylab.org/about/.

Coleman, R. and Hodges, J. (2023) Available at: https://colemanhodges.com.

Donald, M. (2016) 'The performance "apparatus": Performance and documentation as eco-logical practice', *Green Letters,* 20(3), pp. 251–269.

Edensor, T. and Lorimer, H. (2015) 'Landscape at the speed of light: Darkness and illumination in motion', *Geografiska Annaler, Series B, Human Geography*, pp. 1–16.

Environmental Arts Festival, Scotland. (2015) Available at: www.eaf.scot/our-story/.

Hadrians Wall 1900 festival. (2022) Available at: https://1900.hadrianswallcountry.co.uk/events/northumberland-dark-skies-exhibition/.

Hogg, B. (2018) 'Weathering: Perspectives on the Northumbrian landscape through sound art and musical improvisation', *Landscape Research*, 43(2), pp. 237–247.

Ihde, D. (1993) *Postphenomenology: Essays in the postmodern context.* Evanston, IL: North Western University Press.

Ingold, T. (2011) *Being alive: Essays on movement, knowledge and description.* London: Routledge.

Kwon, M. (1997) 'One place after another: Notes on site specificity', *October*, 80(Spring), pp. 85–110.

Massey, D. (1994) *Space, place and gender.* Cambridge: Polity Press.

McClean, S. (2017) Available at: https://acidted.wordpress.com/2017/06/19/incidental-music-from-the-dark-outside-by-stuart-mclean/.

McMullen, T. (2017) Available at: www.alphr.com/art/1006924/an-off-grid-festival-wants-you-to-experience-true-darkness/.

Ratcliffe, H. and Smith, A. (2009) 'Introduction to Warr', in Warr, T., Butler, D. and La Frenais, R. (eds.) *Setting the fell on fire—allenheads contemporary arts: Contemporary art in a rural context.* Sunderland: Art Editions North.

Richards, J. and Shaw, T. (2022) 'Improvisation through Performance-installation', *Organised Sound*, 27(2), pp. 144–155. https//doi.org/10.1017/S1355771821000583.

Ross, L. (1607) Available at: https://maps.nls.uk/atlas/blaeu/browse/932.

Shaw, T. (2023) Available at: https://tim-shaw.info/about/.

Shaw, T., Piquemal, S. and Bowers, J. (2015) '*Fields:* An exploration into the use of mobile devices as a medium for sound diffusion', Proceedings of the International Conference, New Interfaces for Musical Expression, pp. 281–284.

Smith, I. (2021) *Northumberland gazette.* Available at: www.northumberlandgazette.co.uk/arts-and-culture/art/arts-spectacular-taking-advantage-of-dark-skies-planned-for-kielder-3434591.

Turner, L. (2013) Available at: https://thequietus.com/articles/13340-the-dark-outside-review.

Van Den Boogaard, O. (2014) 'In search of Stanley Brouwn', *Frieze*, 161, reprinted. Available at: www.frieze.com/article/search-stanley-brouwn.

Velvick, L. (2016) 'Radio activity', *Art Monthly*, 393, p. 32.

Voegelin, S. (2010) *Listening to noise and silence: Towards a philosophy of sound art.* London: Bloomsbury Publishing.

Voegelin, S. (2013) 'Sonic possible worlds', *Leonardo Music Journal*, 23, p. 89. https://doi.org/10.1162/LMJ_a_00168.

Yezbik, J. (2016) A conversation with Mark Peter Wright. *Sensate Journal.* Available at: http://sensatejournal.com/wp-content/uploads/2016/08/MPW_interview_FINAL31.pdf.

4   The transparency of night

Louise Beer

Introduction

Through my artistic and curatorial practice, I see my life experience divided into two distinct parts, light and dark. Light is the first half, living under the dark skies of Aotearoa New Zealand, and being able to "see" the Earth as a planetary landscape. The second is characterised by more heavily light-polluted skies in towns and cities in England, where a cosmic relationship is harder to grasp. Without both experiences of skyscape, dark skies may not have become a central thematic strand running through my practice. I believe that the increase of light pollution is changing our fundamental understanding of being part of Earth's ecosystem and part of the cosmos. I believe that as our view of the night sky is continuously diminished through this pollution, we may lose perspective of Earth's cosmic significance and that of the climate crisis. In my artistic practice I use installation, moving image, photography, audience participation, and sound to tell the story of our changing relationship to the night sky and our planet (Figure 4.1). I create works that are inspired by science and which offer a way to see through the brightly lit skies to experience the sense of wonder that I feel about Earth and our cosmos.

My childhood in Aotearoa New Zealand involved spending time in many different landscapes. The Hopkins Valley, pictured in Figure 4.1, is the place that I return to in my mind when trying to envisage Earth beyond the bounds of my lifetime.

Being in imposing places such as this helped me to see the foundations of Earth as ever changing, and that our presence is a mere fragment of its history and its future. The dark skies were as much part of the landscape as the tussock covered hills and, unbeknown to me at the time, some of the darkest in the world. This is still the case as Megan Eaves (2023) notes,

> While globally the Milky Way is now invisible to a third of humanity, including most of Europe and North America, in New Zealand it is still visible from 96.5% of the landmass. Almost everywhere in the country is a 30-minute drive from a Dark Sky Place, including the centre of the largest city, Auckland.

DOI: 10.4324/9781003408444-6
This chapter has been made available under a CC-BY-NC-ND 4.0 license

Figure 4.1 Louise Beer, Hopkins Valley, Aotearoa New Zealand.

One experience that connects us to our human and planetary history is the vision of starlight and planets we see as we spin into our own shadow every night. These cosmic bodies hang in our skies all day, only to reveal their magnificence as the light dips from our landscapes. We are losing this view, experienced by our ancestors, at an ever-increasing rate. With each generation, humanity will understand what it is to live under a starlit sky less and less, and our sense of being part of a cosmic system will erode as well. Through five different projects, I discuss the ways in which my experience has been affected by my access to dark skies, exploring themes of loss and grief, the climate crisis, and the deep time spans of Earth. Each project has started with my fundamental connection to the night sky, and each reflects my changing understanding of life, death, darkness, and light.

Gathering light

As a child, I would look out from my father's garden, on top of a small hill which ran along the Otago Peninsula, towards the distant city lights and up at the night sky. Our small city was framed by this cosmic backdrop, igniting all sorts of questions in my mind. Venturing into my father's house and finding scale maps of the solar system, galaxy, and universe imprinted an endless dark space in my mind, helping me to see our landscapes as belonging to an ancient Earth and human beings as ephemeral guardians. As I would close my eyes at night-time, the scale of these places in the cosmos would follow me into my dreams and still today act

as the framework for which I understand the nature of reality. Without this enduring presence of the night, I may have never deeply considered our place in the
universe.

One poignant moment I had as a child was trying to find out what possible
surfaces there could be at the edge of the universe—I imagined a wall, only to
be presented with the answer that there would have to be something on the other
side. The ethereality of perspective that this expansive, never-ending darkness gave
would be of significant comfort many years later, after my dear father passed away.
Deeply feeling the dependable presence of the night-time stars and the moon has
helped me to think of our family's loss from a cosmic vantage point. From my
home in Margate, on the eastern tip of Kent, I see the waves to and fro in the day
and the relatively dark skies at night, reminding me of the lifecycles within all
ecosystems, whether a garden, a mountain range, a planet, or a galaxy. The NASA
images on my walls urge me to think about how the parts of our bodies came from
elements made in the cosmos and will continue to come together in different ways
for the rest of time.

I feel that processing the loss of my father would have been even more incomprehensible if I was not able to look towards the skies and see this framework
with my own eyes, to see the infinite expanse above me, helping me to understand
this giant system of life, death, and regeneration. Looking into the stars helps me
to understand just how privileged I was to share this infinitely small moment of
cosmic history with him.

In 2023 I was commissioned to create an artwork for Derby Cathedral and FOR-
MAT Photography Festival that in some way responded to the Christian practice of
Lent. It was approaching two years since my father had passed away, and it offered
an opportunity to reflect on my loss through my practice. I created a 36-minute
audio visual installation titled *Gathering Light*, using 40 images which I photo-
graphed daily to mirror the 40 days of Lent (Figure 4.2). These images included
sunrises and morning, and night-time views of the moon. The work explores the
rolling feeling of grief that I experience and examines how my proximity to the
coastline, to relatively dark skies and wide daytime skies help me to place my
father's passing within a huge planetary and cosmic cycle.

By panning slowly across each of the 40 images, the installation creates a
sense of a planet *eternally* spinning, suspended in darkness. As an image dips
and fades from one to another, the low rumbling of the sound piece ruptures into
a penetrating crash, gesturing to the vacillating intensity of grief, against the constant rotation of a planet. The work was located in St Katharine's Chapel, a dark
space beneath the cathedral that was accessed by stairs and a corridor. The video
appeared as a three-dimensional sphere due to a vignette placed around the projected light, before it reached the suspended disc. The consistent rhythm of Earth's
daily light and dark cycle are enmeshed in the piece. A consciousness of this vast
eternal system can offer the stability of a constant heartbeat in a time of turbu-
lence. I received many private messages from the audience of this piece, some of
whom shared their experiences of loss with me. The continual momentum of the

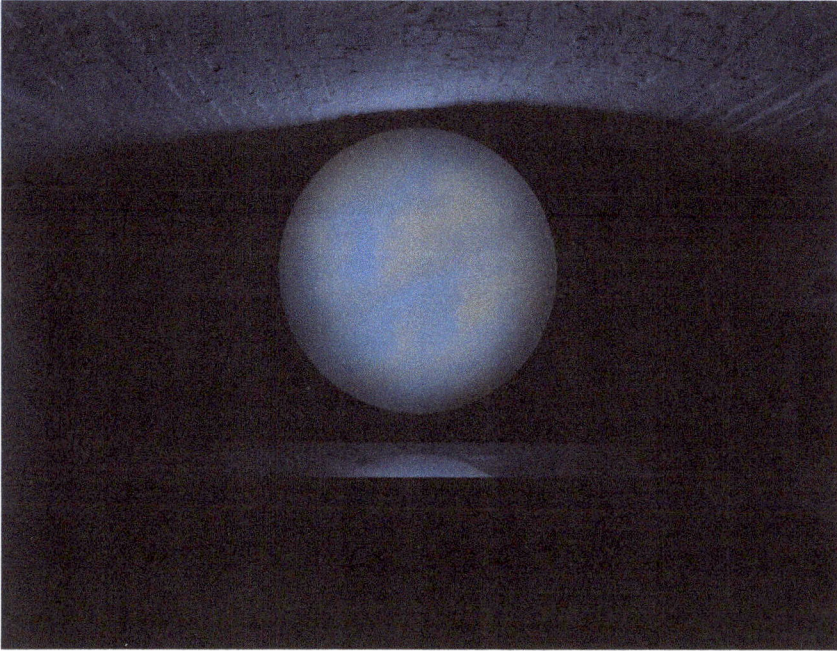

Figure 4.2 Louise Beer, *Gathering Light*, installation view, Derby Cathedral, 2023.

installation offered space to contemplate one's own perspective of life, death, and their relationship to the universe.

The moon, hanging heavy in our skies, is the same moon that my father looked up at all through his life, as did our ancestors 100,000 years ago, and the same moon that reflected light onto the landscapes of our world before there was ever life to see it. Making *Gathering Light* and talking to artists and astronomers about grief and dark skies allowed me to understand how having access to the night sky can be extremely important at different stages and moments of one's life. I have always sought out the night sky to help contextualise issues that I have faced, to feel a sense of being dwarfed by the scale of our universe, but never have I felt the warmth of the stars so clearly as I have in the years after losing my father. Underneath the lights of brightly lit spaces, we cloud over this opportunity to contemplate our place at the end of a long line of ancestors and to see ourselves as part of the same continuous cycle of life and death that exists in our ecosystems, our solar system, and our universe.

A memory of darkness

In 2020, during a residency at the Arts Centre Te Matatiki Toi Ora, I spent several days at Hinewai Reserve in Aotearoa, a nature reserve located on the extinct volcanic outcrop near to Ōtautahi Christchurch in the South Island. I was the only person staying in the huts on the reserve, which is located in a crevice between two hills with the Pacific Ocean washing over the shore below. I spent the daylight hours deeply listening and recording the familiar sounds of the birds and forests through different microphones.

As the light dimmed, the darkness seemed to fall from the sky, gently landing like a veil covering the landscape as the stars and planets began to reveal themselves. The chorus of bird song quietened, and in its place, wind rustled the nearby flax bushes, occasionally joined by the call of a ruru, or morepork owl. The universe seemed to be suspended just at the top of our atmosphere, and I was reminded of how the origins of all of the parts of my own mind, eyes, and ears came from the same distant stars that enabled a ruru's call to resonate through the forest. Whilst gazing up through the darkness, I felt I was "facing the universe". I became consumed by sadness as the stars amplified the sorrow that I felt for the non-human animals sharing the changing light. After our thousands of years of knowledge building and ability to partly understand the time and circumstances it has taken to shape our world, we still collectively tread over their bodies and their homes at our own convenience. This bitter sense of clarity collided with the feeling of wonderment I felt about the rarity of our pale blue dot and its inhabitants. The ancient stars seemed to sit with great stillness in the sky as observers of our delusions.

My experiences of being alone and "facing the night" have been unexpectedly perspective shifting. At Hinewai, far from mobile phone reception, there was little way to break the spell and take myself from one reality into another, to find an equilibrium again. Having a cosmic spectacle as part of our landscapes can act as a perpetual reminder that we are part of a huge, primordial system, and so are the species that we share these landscapes with. McIntosh and Marques (2017) note that "Māori believe that the starting point for wellbeing is a strong sense of interconnectedness with the landscape". I believe that the transparent skies of Hinewai helped me to *face* the gravity of the world's environmental emergency and to further *see* the cosmic significance of each individual non-human residing on our planet. Dark skies can help us to contextualise our own existence as well as the existence of all of Earth's species. They can help us to understand that all of the life we exist amongst came from the same cosmic beginnings, and disclose the privilege it is to share this world with them.

Mark Tredinnick (2008, p. 151) writes,

Mankind has looked to the sky at night since about the beginning; he has found there patterns and stories that explain to him not only where he is and where he should go, but who he is and how small and yet how divine under so much heaven. The night sky is where one fathoms, among other things, one's self.

In learning about ourselves, we are learning about the historical deep time of our planet and its ecosystems. The radiant stars can help us to understand the

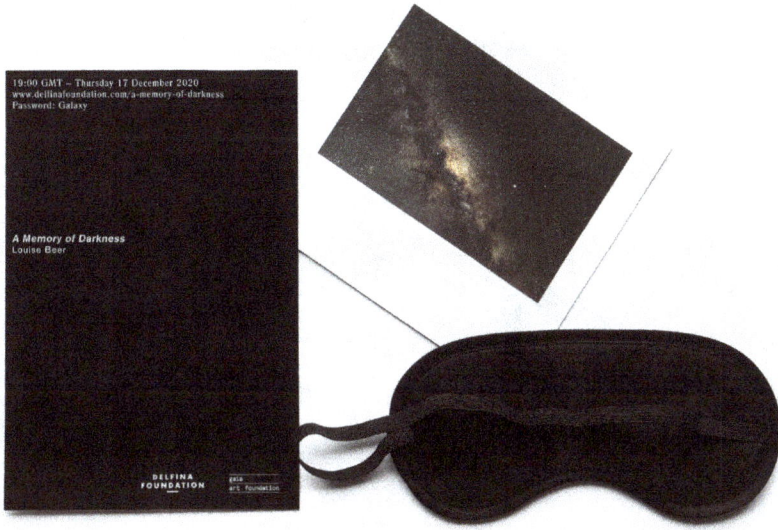

Figure 4.3 Louise Beer, *A Memory of Darkness*: the contents of the parcel received by recipients before the event.

miraculous nature of not only our lives, but the lives of each plant or animal that has come before us. The contrast of dark skies can offer a sense of unity with our non-human companions. I created a sound piece from the field recordings taken in Hinewai that sonically explores the depth of remorse I feel for the non-human animals whose short lifespans we are making ever more difficult without consideration for their right to experience life in the universe. The manipulated birdsong sounds end in a crescendo of grief, as though the inhabitants of the landscape are crying out. This work was the outcome of a UK Associateship I did with Delfina Foundation and Gaia Art Foundation. Their extensive international audiences allowed me to invite people from across the world to listen to the sound piece through a live online broadcast. Those that had registered for the event received a printed photograph of the night sky from Hinewai Reserve, instructions on how to join the event, and an eye mask to wear during the broadcast (Figure 4.3). The eye mask was a symbol of our increasing disconnection to the stars and a way to encourage immersion in the sound. After the event was over, I asked the audience to share their experiences of seeing stars through an online submission form. This piece has also been shown in a gallery setting using wireless headphones (Figure 4.4).

Figure 4.4   Louise Beer, *A Memory of Darkness*: installation view, Bright Island Studio.

Earth reflections

In 2021–2022, I was artist in residence for the Art Tech Nature Culture List, as part of an EU Horizon 2020 research project called CreaTures. The residency was based online, and the objective was to work with the more than 450 people worldwide, of different professions, on the Art Tech Nature Culture mailing list to explore our environment through different methods. I created a series of activities for members to participate in, focusing on our connections to the night sky. For one activity, Earth Reflections, I asked the members to spend time looking at an ecosystem of some sort, whether a tree, a garden, or a river's edge, and to draw or write down what they felt about the climate crisis (Figure 4.5).

The second part of the activity invited the members to look into the night sky wherever they were, and again think about the climate crisis and draw or write down how they felt, before posting the anonymous postcards to me (Figure 4.6). For the first part of the activity, one member described the feeling of "helplessness and anger" at our negligence and disregard of non-human animals and shared environments. For the second part of the activity, the same member described how in the vast emptiness of space, the climate crisis becomes even more important and more disturbing. I found these responses particularly interesting because of the inclusion of "time" in their drawing and text response, not only to indicate urgency but also perhaps to incorporate much wider timescales in our thinking.

Another member said that because she lived under such light-polluted skies in London, she did not feel that she could respond to the activity. Having lived

Figure 4.5   Louise Beer, Earth Reflections I, member response.

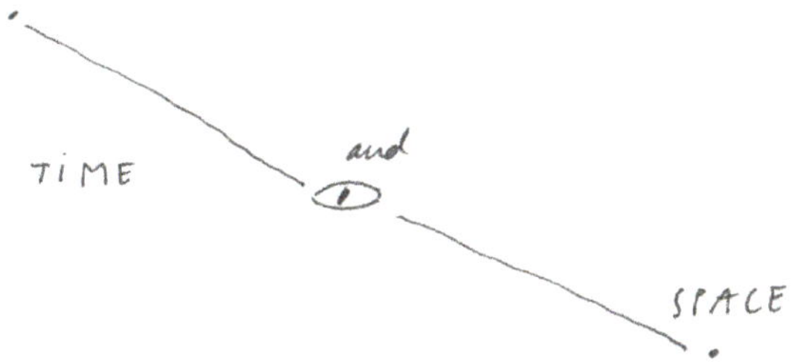

Figure 4.6   Louise Beer, Earth Reflections II, member response.

in London myself for many years, I understood this phenomenon, but the comment left a lasting impression on me. In some parts of the world, it seems that even trying to grasp a relationship with the night sky is now out of reach. I have encountered artists on residencies who had never seen the Milky Way before, and I wonder how different my understanding of existence would be if the stars did not form the foundation of my thinking. William Fox (2008, p. 164) expresses similar sentiments:

> I know of no other place to find our deepest mind than in the night, in the great desert all about us that becomes visible as we pass into shadow and the stars appear to define for us its depth. My fear is not of the night anymore, but of losing it, because then we would have no way to know our place.

I feel an immense privilege to have been born into a place and at a time which allowed me to learn about Earth with a view of the cosmos behind the landscape. Johan Eklöf (2022, p. 152) contextualises this, writing,

> Only one of five people in Europe can see the Milky Way on a daily basis, and in North America and Europe, nearly everyone, 99 per cent, lives under a sky affected by artificial light. Few people know real darkness or what a starry sky looks like. It's almost impossible to imagine the night that was commonplace for mankind only a couple of generations ago.

The night sky has enabled me to see the landscape as an ever changing, timeless land that has reflected light back into the darkness since it came into being 4.5 billion years ago. This philosophical framework has meant that all through my life I have sought out a relationship with the night sky. It is this structure through which I understand our world.

After living in London for 15 years, I moved to Margate, where I have a view of the coastline from my living room window. The rhythms of nature are evident everywhere and my awareness of our solar system and our orbit around the sun is magnified by our star coming into view from this window in April and May as it sets beyond the North Sea. My perspective has changed from looking outwards from Earth into space whilst living in a huge city, to incorporating Earth in my view of the cosmos while living by the coastline.

Earth as a planetary landscape

For five weeks, I was an Artist in Residence in a small village called Chiusure, Italy, on the Amant Siena Residency, during the summer of 2021. The village is perched on top of the incredible white, radiant badlands which appear like imagined mountain ridges on a historical drawing of the moon. The light of the moon illuminates the valleys at night-time, and I spent many nights away from the village taking photographs of them.

One evening, under a blue moon, I walked along the dirt road between two of the valleys, photographing different areas of the badlands. I watched the moon rise over the village and immersed myself in the soundscape, listening out for other creatures. As I was standing on the road, taking long exposures, I heard what sounded like a person in the distance approaching on the gravel. I kept listening as the sound got closer; I remained completely still. Through the darkness, I saw a small, dark shape moving towards me, having no idea that I was there. It was a wild boar swiftly walking along the road under the illumination of the full moon. I shouted "Hello!" so it knew that I was there, and it turned and ran back down the road. I had been warned that the boars could be dangerous, but this was a magical encounter with another of Earth's creatures, spending a moment of its life, at that instant in the history of the cosmos, exploring the same road that I was.

The otherworldly terrain around Chiusure inspired me to photographically explore how it might be possible to visualise Earth as an alien planet. *The Nearest Star to Earth is the Sun* (Figure 4.7) is reminiscent of the Martian sunset photographs taken by the NASA Curiosity rover in 2015. The stars in Tuscany were the brightest I have encountered in Europe and some nights I would stay sitting on the edge of the cliff for many hours, thinking about how Earth was moving through the solar system and around our galaxy and how we can try to imagine the time that was required to

Figure 4.7   Louise Beer, *The Nearest Star to Earth is the Sun*, Chiusure, Italy, 2021.

create every molecule that makes up every part of the ecosystem. As Thomas Halliday (2022, p. 5) writes, "The aggregations of species that produce a feeling of place also provide a sense of time". Not only is it a challenge to think about how unique life on Earth is in all of the endless universe but that our experience of it and what feels familiar is connected to this moment of Earth's inconceivably long history and future.

I am interested in how we can think about the environments we know as planetary landscapes that have risen and fallen over billions of years and about the number of sunsets and sunrises that have occurred, where the geology beneath the stratosphere is changing, and life alongside it. Without these developments, we would not have the same landscapes and forms of life that we recognise today. If we think about the formation of this landscape in the context of the galaxy and universe, how does it challenge our thinking about the importance of the lifeforms and environments that we live amongst? Does it challenge our human-centric view of the world? Guy Kahane (2013, p. 764) states, "If we are alone in the universe, the only thing of value, then this gives our continuing existence, and our efforts to avert disaster, a cosmic urgency, on top of whatever self-interested, anthropocentric reasons we have to stay around".

I created a body of work titled *Earth as a Planetary Landscape* by layering and manipulating photography of sunsets, sunrises, moonlit landscapes, and night sky photographs taken in and around the badlands in response to these encounters with the night sky in Chiusure. The series of works tell a narrative story of Earth's long history and the impact of the climate crisis, both visually and through descriptive titles. The first piece in the series, *Through Time and Distance We Cannot Understand*, depicts a life-giving interstellar object. The second piece, *Eternally Spinning through Darkness* (Figure 4.8), and the subsequent piece, *A Twist, a Turn, under the Shadow of the Moon*, give a sense of the longue durée, as Earth slowly changed into the world we know today. *Apogee Earth* (Figure 4.9) portrays our world just before the devastating effects of the Anthropocene, as also suggested in *Blinded by Light, a Gathering Collapse*. The final image in the series displays a still blue world underwater, perhaps with little or no life at all, titled *Everything is Forgotten*. The pace of changes to Earth increases in speed across the series, reflecting the rate of the changes induced by the climate crisis, in contrast to the long and slow developments over geologic time. The use of multiple panels in the first three pieces suggest the coalescence of different elements and the single panelled pieces suggest a formed world, a home for life. The series seeks to provoke a sense of the vast time needed to form the world in which we inhabit and the quickening, sudden, and devastating impact of the climate crisis. Martin Rees (2003, p. 157) argues, "The odds could be so heavily stacked against the emergence (and survival) of complex life that Earth is the unique abode of conscious intelligence in our entire Galaxy. Our fate would then have truly cosmic resonance". This series of work endeavours to capture the significance of each outcome of evolution thus far, not only human life but of myriad non-humans with whom we have together created the complex puzzle of our ecosystems.

Figure 4.8 Louise Beer, *Earth as a Planetary Landscape/Eternally Spinning through Darkness*.

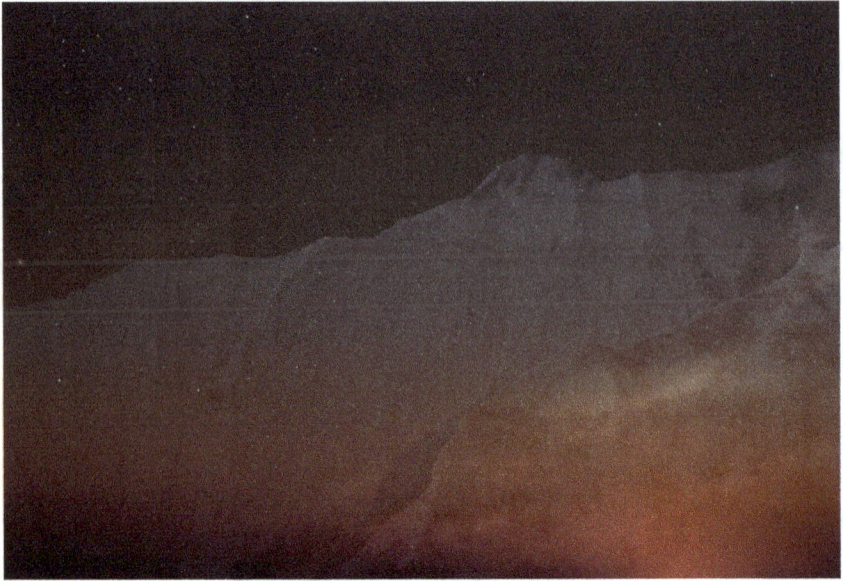

Figure 4.9 Louise Beer, *Earth as a Planetary Landscape/Apogee Earth.*

Cosmic contemplation of a planetary catastrophe

I am interested in how scientists who have a deeper understanding of time and distance, and the potential rarity of life in the vast cosmos, can live within what seem like dual realities. How is it possible to know the universe in one sense and to see humanity's wilful destruction at the same time? This is even more astonishing given how unusual our ecosystem appears to be, for, as Martin Rees (2003, p. 168), argues " . . . the emergence of intelligence may require such an improbable chain of events that it is unique to our Earth. It may simply not have occurred anywhere else, not around even one of the trillion billion other stars within range of our telescopes". In 2022, I asked astronomers and dark sky workers to reflect on my question, inviting responses which examine these two temporalities, as earthly climatic change shifts from occurring across thousands, or millions of years, to rapid changes in our lifetimes:

> How does your knowledge of astronomy, the age of the Earth, the age of the universe and the rarity of life (as far as we know so far) frame your understanding of the climate crisis? Do you think about the significance of the climate crisis from a cosmic perspective?

Beatriz García, PhD in astronomy, member of Pierre Auger Observatory and QUBIC Collaborations, and member of the National Council of Scientific and Technological Research (CONICET), responded as follows (Beer, 2022):

Life is something precious, which appears at the end of an evolutionary chain that took about 14 billion years (yes! the life of the universe), which may exist in other places, but we still do not know it; what is real, is that in the third planet in order of distance from a star called the Sun . . . there is life and this is the only possible world for life as we know, from unicellular to the most complex beings, those that even manipulate matter by the one that took those 14000 million years to be able to replicate. Climate change is a central issue that did not begin in the 21st century, of course, but for which we human beings in this century bear absolute responsibility. Are we really willing for something as significant for the universe as life to be extinguished because we simply don't care about anything? It is probably difficult to reverse some processes, the catalytic power of men and women to accelerate some natural processes and introduce unsolvable problems is enormous . . . however, I like to think that we are in time to slow down at least the inexorable line of time stopping the soil, air, water and sky contamination, using a more rational amount of energy, consuming just what we really need, assuring the good distribution of the resources, educating the citizens to protect the environment, following the scientific recommendations, and this leads me to remember a great phrase of probably the most famous Argentine writer, Jorge Luis Borges, which is very appropriate for closing: "The future is not what is going to happen, but what we are going to do.

Dark skies can teach us the significance of every moment, of every evolutionary pathway, of every species of life. It is imperative that we protect our existing dark skies and reduce light pollution so that we can be more conscious of being inside Earth's ecosystems, not separate from them. In 2023, astronomer Dr Lindsay DeMarchi responded to my question, exploring similar ideas and compounding my feeling of privilege by also having a profound connection to the night sky (Beer, 2023):

Take a moment to consider that we, alive with breath, standing on a planet teeming with vitality, yelling into a dark, quiet, cold expanse, is incredible. We are conscious. We have a dedicated rock with an atmosphere catered to our breath, and a ground that trades seeds for food. Emerson once said, "I am a transparent eyeball". He sees everything but he is also a seamless part of everything, and he cannot see himself. When you do see yourself, you understand that you are already a part of the meaning. That is why, as a scientist and as a sentient speck of the universe, I am dedicated to the stewardship and protection of this planet and the quality of its life.

Cosmic time periods feel eternally out of our psychological reach. Looking back through Earth's history can help us to understand not only the unthinkably small chance that life occurs at all but that the events which occurred have created an environment that has allowed life like *ours* to exist, as Thomas Halliday (2022, pp. 108–109) emphasises.

In the complex game that is an ecosystem, every player is connected to some, but not all, others, a web not just of food but of competition, of who lives where, of light and shade, and of internal disputes within species. Extinction bursts through that web, breaking connections and threatening its integrity. Sever one strand, and it wavers, reshapes, but survives. Tear another, and it will still hold. Over long periods, repairs are made as species adapt, and new balances are reached, new associations made. If enough strands are broken at once, the web will collapse, drifting in the breeze, and the world will have to make do with what little remains.

Conclusion

Does looking out to the stars from our blue world, listening to the soundscape sung and created by all the other elements of the present ecosystem, each with their own momentous cosmic journey, travelling through the solar system with us, not feel like looking out at a sublime, uninhabitable landscape from the vital warmth of a womb? *The Overview Effect*, by Frank White (1998), has deeply inspired my practice. I cannot imagine anything more perspective-shifting than seeing our blue world gliding through our solar system. However, seeing through a dark sky to elements of the cosmos, suspended in a black sea above our world, must be a close second place when accompanied by the call of a distant ruru or the quiet footsteps of a wild boar. Through all the years of warnings and statistics, the destruction of our planet rages on and our human and non-human companions continue to pay a heavy price. Artworks that explore our relationship to the cosmos can help us to reconnect with something that we may have already forgotten is missing and re-evaluate how we each understand the cosmic significance of our world and all its inhabitants. Artists can offer an emotional reflection on science and provide an open window to find the magnificence of the cosmos and our place within it once again. By offering an opportunity for an audience to consider their own changing relationship with the night, they may feel empowered to protect it.

If our atmosphere was not transparent or if our eyes had not developed in a way that could detect the faint light of distant giants, would our minds be so full of wonder and inquisition? Lord Martin Rees (2003, p. 8) declares, "What happens here on Earth, in this century, could conceivably make the difference between a near eternity filled with ever more complex and subtle forms of life and one filled with nothing but base matter".

We are merely at a point in the continuum on our planet in which time will pass for a colossal period before its eventual end. In all the vastness of the velvet black sky, it is our home which is glorious, generous, and forgiving. Just as we are each the result of billions of years of changes and adaptations, so are all the other elements of each ecosystem we inhabit. If only we could gaze into the eyes of the creatures that we share this moment of Earth's cosmic history with and understand their evolutionary path, their right to exist now and in the future. Dark skies can plant a seed of curiosity for our historical deep time and our deep time future and help us to understand our connection to both our familiar and distant ancestors who shared this same sky. As we are all biologically affected by artificial light, I believe that we are also philosophically influenced.

References

Beer, L. (2022) Email conversation with Beatriz Garcia, 22 April.

Beer, L. (2023) Email conversation with Lindsay DeMarchi, 5 May.

Eaves, M. (2023) 'Land of the long dark night', *Nightscape*, [online] March, p. 13. Available at: www.darksky.org/wp-content/uploads/2023/03/Nightscape-111-Mar-2023.pdf (Accessed: 20 March 2023).

Eklöf, J. (2022) *The darkness manifesto*. London: Penguin Vintage.

Fox, W. (2008) 'Night in mind', in Bogard, P. (ed.) *Let there be night*. Reno, NV: University of Nevada Press, pp. 159–164.

Halliday, T. (2022) *Otherlands*. Kindle edn. London: Penguin.

Kahane, G. (2013) 'Our Cosmic Insignificance', *Noûs*, [online] 48(4), pp. 745–772. https://doi.org/10.1111/nous.12030.

McIntosh, J. and Marques, B. (2017) 'Mātauranga Māori and therapeutic landscapes', *www.academia.edu*. [online] Available at: www.academia.edu/35808248/M%C4%81tauranga_M%C4%81ori_and_Therapeutic_Landscapes?email_work_card=title(Accessed:3 May 2023).

Rees, M. (2003) *Our final century*. London: Heinemann.

Tredinnick, M. (2008) 'Original country', in Bogard, P. (ed.) *Let there be night*. Reno, NV: University of Nevada Press, pp. 150–158.

White, F. (1998) *The overview effect: Space exploration and human evolution*. Reston, VA: American Institute of Aeronautics and Astronautics.

Part 3

Sensing dark landscapes

5 Nightfalling

Dancing in the dark as an artistic practice

Ellen Jeffrey

Momentarily, the curvatures and contours of this tilted planet slide into the darkness of its own shadow. "Nightfall" describes the journey of a place towards that shadow, of a place turning into shadow. It is a time of transition, of shift and otherness. It is a time of heightened sensitivity towards the planetary motion of this rotating earth. It is, in many ways, a most suitable time for dancing. This chapter explores what we can learn of the night by dancing in the dark: what we can encounter of the nocturnal world by venturing out and moving with it. In what follows, I consider whether by conserving night's darkness we might also conserve ways of moving through the world—of meandering, wandering, and slowing down—and how these ways of moving might be essential to our encountering and understanding of the more-than-human environments we live with.

My interest in moving outdoors at night began in Finland, and it began with walking. In August 2011, I moved to Helsinki to study at the city's University of the Arts. For two years I lived in Kumpula—a small town on the northern outskirts of Helsinki, a place surrounded by pine woodland. During my first winter there, the nights lengthening and my sleep fragmenting, I began walking through the woodland at night. Following the meandering pathways that circled through the trees became a way of passing the night's many sleepless hours. As I walked, I observed how the night's darkness shifted with the seasons: how the woodland's deep dark of late autumn could be transformed within hours by a heavy snowfall that illuminated the paths and set the whole woodland aglow. The snow transformed not just the landscape but my movement within it: my walk shifted from a quiet stroll into a repetitive, crunching tread—each step announcing my presence in that place, forming a soundscape to accompany my walks which would last until late spring. And then, summer nights: nights that were in fact not night at all but instead a long turquoise twilight that made my walks in the early hours feel to be a slipway between the after-dusk and almost-dawn. Only occasionally would I catch sight or sound of someone else in the distance, and one of us would shift to take another path, keen to sustain the solitude that had been so carefully cultivated by the night hours.

Kumpula's nights were different to any I had known in England. They were different in depth, in quality. Kumpula is seven degrees further north from where I grew up in North Yorkshire, where the landscape of agricultural farming means skies are vast, open, and with an earth-carved horizon. I did not know then, as a child, how lucky

DOI: 10.4324/9781003408444-8
This chapter has been made available under a CC-BY-NC-ND 4.0 license

I was to know night as darkness, to know night to be a spectrum of blues and greys shaped by the weather, seasons, and land. I was lucky to know what night felt like and to feel the exhilaration of being out at night, an exhilaration characterised as much by fear as it was by freedom. Those two years I spent nightwalking in Finland reminded me of what it was to be out in the night's dark, and what it was to be immersed in the transitory, gradual process of its darkening. When I left Finland upon completion of my studies and returned to London, I returned to the orange hues of city nights: to a starless stasis that stretched across Deptford's high street and endured the year through. For two years, I continued to walk in London at night. Working early shifts in cafés and department stores, I would re-route my journey to seek out the strange night-time spaciousness of Trafalgar Square, or the shaft of natural darkness spilling out across Deptford's muddy creek. On days off I would lie wakeful in those pre-dawn hours that had become my work hours, before giving up on sleep and heading out to walk along the edge of the Thames' southern shore, listening to the calming slosh of its waters, aglow in the electric overspill of the city's night.

Like London's other nightwalkers, I moved always with a sense of purpose and direction even if, in truth, I had neither. These nights were not without incidents of strangeness or fear: I missed the freedom to meander, and to meander unseen. In moving to London from Kumpula, I felt the loss of the night's dark and the varied spectrum of its processual darkening. But I also felt the loss of the ways in which night's darkness made the world obscure to me and, in turn, obscured me to the world. I began to read articles that used data and measurements to address the increasing spread of light pollution across the surface of the earth, describing in numbers and figures the not-so-gradual diminishment of natural darkness. Yet these did not articulate the particular experiences of moving and encountering that I had of night-time, experiences that I felt to be reliant *on* the dark and diminishing *with* the dark, experiences that I felt could not be adequately articulated through words but perhaps needed to be comprehended as—and through—movement. After almost two years of nightwalking and night-working in London, I chose not to remain within the receding dark of its cityscape and instead seek out the potential ways of moving that were made possible by night's darkness. I wanted to work towards an understanding of moving-with-night—a practice of moving-with-night—that did not simply rely upon darkness but insisted upon its necessity.

***

This kindling of interest in moving through night-time places laid the foundations for what became my practice-based research at Lancaster University. Alongside my café shifts in London, I had worked sporadically as a professional dancer—dancing in theatres, galleries, and studios, in the UK and abroad: sleeping in hotels, on couches, in sub-let rooms and shared beds. In commencing this period of artistic research, I sought to blur my wandering with my dancing, and establish a practice through which to explore how night-time darkness shapes the ways in which we move, and the ways in which we kinaesthetically relate to a place at night. Over a period of four years, I worked in just one site, Grubbins Wood, an area of ancient woodland situated on the southern edge of the Lake District National Park and managed by Cumbria Wildlife Trust. The research took place specifically at nightfall, during those hours of processual darkening that are at once almost-night and not-yet-night. The practice created a series of workshops, one for local participants and another for professional dancers, and generated two night-time choreographies that took place over the duration of nightfall: *On The Patterns We Gaze* (2019) and *On The Traces We Carry* (2021).

As a dancer, before I began dancing in the dark in the wood, I believed that the skills I had acquired through training—the ways of moving that I had learnt as a dancer—belonged to me, that I held that knowledge within my body, to be executed at will. It was not until I began practising in the wood that I began to recognise that my way of dancing was entirely facilitated by the spaces in which I danced and by the similarities of those spaces: by their flat floors, the warm temperature, their consistent level of light. I began to realise that my dancing did not exist *in* the space but rather only ever existed *with* and *through* the space. Learning to dance with(in) the dark of Grubbins Wood meant I had to learn to listen—not simply with my ears but to listen kinaesthetically, to listen through my movement as a necessary form of navigation. Having been accustomed to training in indoor studios, where spatial alterations are largely human-made—the shifts in movement by myself and other dancers, the initiation of sound by musicians—then dancing in the wood meant not only attuning to the motions and surfaces of other-than-human beings, but also to how the weather, seasons, and changing level of darkness constantly altered the environment and the inhabitants that composed that environment.

As it turns through the seasons, the woodland becomes the dwelling-place of some and the passing-place of many: amongst them are red wood ants, southern wood ants, Lancastrian whitebeams, hart-tongue ferns, wild garlic, ancient yew trees, tawny owls, bluebells, pheasants, roe deer, ravens, and of course, humans. In becoming a regular night-time visitor to the wood, I was wary of whether my moving body had the capacity to simply be a part of the wood's crepuscular activities without displacing the movement of others. With this in mind, I began working along the wood's more sparsely populated southern edge—here, an immense canopy of darkness is held between the copious branches of yew trees and the deep-cut furrows of the path that winds below them. Even in daylight this path remains somewhat shadowed and muted, the deep green tongues of the ferns luxuriating alone under the yews' canopy—dimming, dampening, quietening. This became the area of woodland I frequented most, inhabiting the path through pace and gesture, patterning my dancing form through the duration of dusk and into night. With time, movement became a sort of additional sense, an alternative way of understanding the night-world of the wood.

In beginning this practice, the most significant adjustment was my relationship to the ground. As a dancer, even when you dance alone, you are always duetting with the ground, with the surface of the earth and the invisible ties of gravity that exist between you. In the wood, the ground is an undulating and folding surface, a living space that sinks and rises, one that is made up of both the burrowing roots of trees and the flourishing stems of saplings. Whereas the surface of a studio floor enables the dancer to find, lose, and re-find a sense of stability, the woodland floor at night seems to create a constant unbalancing. In the dark, the feet sort of seek and discover what the eyes cannot. At nightfall, the ground would determine my movement in such a way that felt that rather than dancing on it, it was moving me and moving through me: or in other words, rather than my body being in motion across the ground, the ground was in motion through my body. To explore this sensation further, I began by simply walking through the wood. As I walked, I held pencil to paper, allowing the alterations and transfer of weight between my feet to feed upwards through my body so that what appeared on paper are tracings of how the crinkling of the wood's ground is felt at night through motion. Of course, the hand is not really tracing the tread of the foot—that alternating rhythm—but rather the flow of the motion, a flow which is determined by the relationship between the weight of the body and the texture of the ground. The jolting lines that come into being on the page are neither straight nor direct—they did not record a sense of direction or destination but rather, to borrow the words of artist Paul Klee, they seem to go out "aimlessly for the sake of the walk" (Manning and de Zegher, 2011, p. 1). When sharing this task within the night-time workshops, participants described a sort of motional drift, their movement "having no end place", "naturally enhanced by gravity and stumbles", "sometimes carried sideways and back", and becoming a way of experiencing a kinaesthetic relationship with the other-than-human environment of the darkening woodland: "movements naturally followed the features of the landscape", "the branches, tree trunks and fallen twigs defined my movements for me", "I felt my way with my feet in the steep rocky ground". The basic pencil tracings that emerge from this practice document are not simply the wanderings of the feet but the "no end place" of a ground-in-motion (these are extracts from workshop participants writings, produced during the series of workshops run on-site, 2018–2019).

In performing this task over the hours of nightfall, my movement transitioned from walking into wandering into dancing. The slowing and meandering that night's darkness does not simply allow for but requires a way of encountering the world that is diminishing in both (human) value and (human and non-human) access. In losing night's darkness, we lose not only ways of moving but ways of moving-with: night creates an environment in which it becomes possible for us to meander, to move aimlessly, to get lost—but it also allows us to follow the ground, to experience ourselves not as individual units that tread upon the surface of the earth but rather as motional bodies that are moved by and with the environments of which we are a part. In the dark, the undulations of the wood's ground act as small, unexpected nudges away from a vertical and stable stance. Just as there is a loss of directness in the formation of my pathway, so too is there a loss of directness in the placement of my weight. As the wood darkens into night, the transfer of my body's weight shifts from being between feet and ground to being between a multitude of surfaces and materialities: spine, pelvis, skull meeting bark, moss, fern, mud. Dance scholar Vera Maletic describes how "every movement is a composite of stabilizing and mobilizing tendencies and, consequently, the act of balancing constantly exists" (Maletic, 1987, p. 198). In dancing in the dark, these "stabilizing and mobilizing tendencies" do not feel to be held in the body alone but emerge through encounters with(in) the fabric of the motional environment. In other words: the movement that forms feels to exist beyond the parameters of my body, as if by dancing in darkness my nervous system has the capacity to interlace with the electric synapses of other inhabitants of that same darkness, reaching-sensing-responding, a web amongst webs. This loss of stability, of verticality, of certainty in where my movement ends and the environment begins—it feels like the night's darkness has the capacity to shift humans sideways, to topple any sense of a hierarchy that has built up between ourselves and the rest of the world. As Jane Bennett observes, "to begin to *experience* the relationship between persons and other materialities more horizontally, is to take a step toward a more ecological sensibility" (Bennett, 2010, p. 10).

Moving in the dark, then, makes felt not only how entangled we are in the world but also how dependent we are. To move in the dark is to be humbled by the dark. When we walk through woodland in the daytime, we can see the impression our bodies make on the ground—we leave footprints, we make pathways through our tread, and we can see very readily the impact of our presence and those that were present before us. What we do not so easily see is how the ground makes its impressions on us—but when we walk in the dark, the ground makes itself felt. In a contemplation of (and, also, as) walking, Tim Ingold describes how "the complex surface of the ground is inextricably caught up in the very process of thinking and knowing" (Ingold, 2015, p. 49). In losing darkness, we lose ways of encountering the ground (ways of encountering the world) which shapes and co-forms the very processes by which we move, think, and imagine. To dance in the dark is to experience not the body's motion meandering along the ground or page but rather the contours and alterations of a ground meandering through the body's motion.

***

In sharing this practice with others, the questions that arise often concern a common topic—vision. People ask how it is possible to dance without seeing or to perform without being seen. During these conversations, I am reminded of certain facets of my professional dance training: studios where motion-detector lights and mirrored walls acted as a constant reminder that I trained not simply to dance but to be seen dancing. In the dark, in the wood, I dance with a sense of having momentarily escaped not only the intense critique of my own eyes flitting across a mirrored reflection but also from feeling that the value of my dancing lay only in its capacity to function as a spectacle. I have trained to dance and to move before bodies that are seeing and stationary: in the moment of performing, I usually experience the choreography's unfolding as being suspended somewhere between my dancing form and the viewer's perspective. Dancing in the wood, the hours of nightfall enable a re-focusing of the potential ways in which human and non-human movement might be both visually and kinaesthetically encountered.

There is a strange vibrancy to this north-westerly edge of land where the woodland dwells. It is almost as if splices of several different landscapes have become wedged alongside one another, their seams no longer discernible as one tumbles into the next: the ferns and mosses unfurling over the shadowed and furrowed earth of Grubbins Wood open out on to two rolling grass fields whose pastures descend gently towards the coarse turf of saltmarsh that fringes the immense sands of Morecambe Bay, which are shifted and silted under the watery weight of the river Kent and the Irish Sea as they meet and merge in the estuary. It is a place of dynamic interplay, not just in the grit and sludge of the terrain but also in the bodies that move through it and call over it—and at night, this interplay inhabits the darkness and cultivates my movement. In this seeping night-space, the wood becomes a world where what is visually perceived no longer equates to clarity and accuracy—imagination anticipates form rather than recognises it. As Gaston Bachelard suggests, "space calls for action, and before action, the imagination is at work" (Bachelard, 2014, p. 34). As I dance in the dark, moving and imagining becomes an entangled way of perceiving the world: the shapes and forms of the things I see gradually blur and merge into one another, becoming something other, something unfamiliar. As nightfall progresses, forms gain movement in my perception as I attempt to comprehend the boundaries of their shape. My body is motion amongst motion, the spatial forms and spatial tensions of roots and branches, footpaths, and flightpaths, both seen and unseen, real and imagined. One workshop participant describes this as the "blurredly blur", nightfall seeping into their vision, "everything becoming the same colour". Over the duration of an hour's practice, my vision loses the clarity of distinct form and instead perceives the potential existence of a relatedness between things, a relatedness that is entangled, seeping, and shifting. In this way, human eyes that open to the night do not see less of the world but more of it: they see the potential for things, bodies, and surfaces to exist beyond the parameters of their form, to be more than what they seem. In the moment of movement, the "blurredly blur" of visual perception becomes kinaesthetically felt. Just as seeing in the dark reveals the ways in which forms shift, merge, smudge, and become-with one another, dancing in the dark makes tangible the ways in which movement spills beyond form, existing always between and beyond material structures. As dance scholar Bojana Cvejić asserts, "the shadow is also another body" (in Bauer, Cvejić and Laermans, 2011, p. 5).

Whilst observing workshop participants dancing in the wood at the very cusp of twilight, the blurring of forms appears to be the perception of nightfall in motion. Writing on-site, I describe how

the face of the furthest dancer is lost to me, and the others are becoming so. Night seems to be a horizontal movement in this place, a creeping-towards rather than a falling-down. The path acts as a measure of the night's smudging, the smudging of forms and details. Things here are becoming two-dimensional, becoming sketches of themselves.

Here, the dancers' presence acts as a constant, accentuating the processual darkening of the environment. This "smudging of form and details" by which things are "becoming sketches of themselves" is the appearance of things beyond the form of their structure. Maurice Merleau-Ponty suggests that by neglecting to paint the outline of things, Paul Cézanne was able to abandon himself to "the chaos of sensations" and "depict matter as it takes on form" (Merleau-Ponty, 1964, p. 13). Perhaps through this practice of dancing in the dark, it becomes possible to consider that at night we perceive—both visually and kinaesthetically—not the loss of form, not a disappearance, but rather the appearance of formation.

In each nightly return to the wood, in each movement re-patterned by darkness, there exists a folding over of relations: one that seeps and sifts not just with the immediate, darkening environment of the wood, but also with each shadow-body of the night before, the season before, the year before. If, as dance scholar Valerie Preston-Dunlop asserts, "space speaks" (1998, p. 175), then it is possible to assume that the tone and cadence of a place can, like a voice, be heard and be recalled. It is only as this practice extends itself through the years—my body re-finding movement in different nights, in different weathers, in different locations of the wood— that I am able to recognise the mistake I make in describing "the dark" as though it exists as a single, consistent state. In the process of returning and re-doing, my body becomes sensitive to all that is altered: how the density and quality of night's darkness shifts with time and with place. Some nights, the wood enters a perpetual state of enchanting stillness and nocturnal brightness, the entangled branches of the yews above patterning the earth below with their moon-shadows, commingling with that of my own dancing form. In these nights, the scale and distance of the wood feels palpable within the slightest of gestures. In others, the intensity of the weather renders the darkening wood unrecognisable, and my whole focus aligns to the contours of a single path and the direction of the wind—branches newly fallen, earth freshly saturated movement drifting and orientation faltering. Just like the movement of our bodies, a night's darkness is an expression of its relatedness to an environment, its relatedness to things: it is a betweenness shaped by the shifting transformations of others, perceptible to our moving selves as, like us, it weaves itself into a place's infinite lines of becoming.

When I dance at night, I have an understanding of myself as a body in motion: and by that, I mean as a form that is capable of mobility and articulation. However, when I dance in—and with—night's darkness, I have an understanding of myself as a body *in* motion: a body immersed within motion, within the motion of the site, within the motion that is nightfall itself.

***

It is 10:15 p.m., −18°C, 400 metres above sea level, and I am lying on the flat-topped plateau of a hill just outside the small town of Longyearbyen, Svalbard, 78° north and stargazing. It has been five days since I last saw the sun, watching its red glow slip beyond the horizon as the plane continued its journey north. Since then, the darkness I so readily associate with night (indeed, define as night) has been constant, so that day and night are differentiated not by alterations of light but by human activity: shops open, clocks turn, shifts begin and end. Lying in the snow, my gazing eyes see a distorted night sky, distorted only because they gaze with the expectancy of the familiar, of identifying constellations in particular relation to one another. But this far north, there is a difference: the patterns are made strange to me by my relocation, and there are far more stars, more patternings inhabiting each constellation than I have ever seen from the wood. The wood feels very far away from this place—there are no trees on Svalbard. As I look, my fellow stargazer lies down next to me and points out Cassiopeia and talks about how the sky appears differently here than from his home in southern Norway. He recalls stargazing with his family, and as I look out, I think of all the other times my eyes have sought these patterns and all the other eyes that are seeking them now, in this time. I think about how the sky is a part of what defines a place. That perhaps these vast distances between people and stars, between stars and stars, is made irrelevant by our patternings that seek to entangle: to entangle by words and stories, by drawings and diagrams, by movement and imaginings. Lying in the snow, I feel so very small and fleeting in this night that is also day, this night that is a space as much as it is a time. As Don Handelman suggests, the night's darkness is one such force which holds us: describing it as "sticky", clinging continuously", it is an entity "from which the person cannot separate" (2005, pp. 253–254). Perhaps more so than any other atmospheric entity, night's darkness simultaneously reveals to us the vast scale of the universe whilst also reminding us of the smallness of our place within it: that we are held by and shaped by the lines of force, the fluxes and flows of an atmosphere—an atmosphere that is at once place-specific yet also entangled within that same darkness which holds the far-off light of the stars.

Here, twilight no longer exists as a transition between light and dark—wakefulness and restfulness—but instead as a few hours of vague and subtle brightening towards the south-east horizon. To me, these hues of the polar twilight are more beautiful than even the northern lights themselves: with each appearance they shift, shaped by the brightness of the moon and the cover of clouds, saturating the deep snow of the arctic with a bluish glow. They fascinate me, and it occurs to me it is the first time I have encountered a night-light such as this—compressed within a few short hours, it is simultaneously dusk and dawn. Writing almost 100 years earlier, Christiane Ritter describes gazing out over this same arctic nightscape: "Whatever we feel, it is not loneliness as we stare into the distance. It is as though we are enclosed in a miraculous world" (Ritter, 2019, p. 126). Ritter's "miraculous world" is not, I think, limited to the polar night but encompasses all nightscapes where the subtle hues of darkness thrive unhindered. Lying above Svalbard's frozen shore, the web of my nervous system is alert to the newness of this night: to the scrape and crunch of arctic snow beneath my shifting weight, to the tingling sharpness of frozen air against my skin, to the strange forms of the treeless hills over which my body clambers, rolls, and lies. Yet it also resonates with the sensation of moving through nightscapes that exist beyond this place and time: the soft slosh of the Thames near Deptford's creek, the cool floor of a dance studio in the dark, and the smell of the sodden, seeping, and vibrant earth under the yews of Grubbins Wood. I lie on the plateau for a few moments more: my eyes re-find Cassiopeia and I trace its form with the tip of my nose, my skull rolling inside my woollen hat and hood, spine following skull, making small, scuffled imprints in the snow.

References

Bachelard, G. (2014 [1964]) *The poetics of space*. London: Penguin Books.

Bauer, B., Cvejić, B. and Laermans, R. (2011) *Hear my voice*. Translated by Griffin Translations. Ghent: New Goff.

Bennett, J. (2010) *Vibrant matter*. Durham and London: Duke University Press.

Handelman, D. (2005) 'Dark soundings—towards a phenomenology of night', *Paideuma, Frobenius Institute Frankfurt*, 51, pp. 247–261.

Ingold, T. (2015) *The life of lines*. Oxon: Routledge.

Maletic, V. (1987) *Body—space—expression*. Berlin: Walter de Gruyter & Co.

Manning, E. and de Zegher, C. (2011) *Violin phase: Anne Teresa De Keersmaeker NY 2011*. New York, NY: Mercatorfonds.

Merleau-Ponty, M. (1964) 'Cezanne's doubt', in Dreyfus, H. L. and Dreyfus, P. A. (eds.) *Sense and non-sense*. Evanston, IL: Northwestern University Press, pp. 9–25.

Preston-Dunlop, V. (2006 [1998]) *Looking at dances*. London: Verve Books.

Ritter, C. (2019 [1954]) *A woman in the polar night*. London: Pushkin Press.

6 Sensing dark places

Creating thick descriptions of nocturnal time and rhythm

Rupert Griffiths, Nick Dunn, and Élisabeth de Bézenac

Introduction

Darkness is relative. It usually refers to the presence of little light rather than its complete absence—from an architectonic horizon silhouetted by urban sky glow or arboreal shadows feeling their way through a moonlit forest to the crisp firmament of a clear and moonless night. Even when confronted with a complete absence of light, deep in a cave or in a windowless room, we experience light—those orbs, shimmers, and sparks floating before us created not by light from our environment but by the perceptual noise of our sensory apparatus. It is this spectrum of dark illumination that this chapter aims to address, asking what value darkness holds and how this can be captured and communicated. It considers how we can create thick descriptions of darkness and the purpose and value of doing so. Rather than approach darkness as in opposition to light, darkness is approached as a nuanced scale that slips from the bright illumination of the midday sun to the deep night when we rest in the shadow of the earth. It will consider the tints and shades of environmental light as one moves through night and day. Equally, it will consider the presence of human sources of illumination, some of which might be thought of as light pollution, although perhaps dark pollution would be a more accurate term.

This chapter discusses three approaches to the experience and description of darkness—walking, photography, and unattended light sensors—and how these approaches intersect and diverge. It will draw from a research project and fieldwork undertaken in Leighton Moss RSPB Nature Reserve in Cumbria's Arnside and Silverdale area of outstanding natural beauty, weaving this into a discussion of the particularities of experience and description that walking, photography, and unattended sensors can bring to our understanding of darkness and its value in specific places. It will then consider how these three approaches might be brought together to create a thick description of the night that is greater than the sum of its parts and how this might help us to become attentive to darkness and its value for both humans and non-humans. Finally, it considers how such descriptions might be useful in terms of informing urban design and policy in ways that are sensitive to the values and challenges associated with darkness.

DOI: 10.4324/9781003408444-9
This chapter has been made available under a CC-BY-NC-ND 4.0 license

The value of light and darkness

Sunlight is the source of energy for almost all life on earth, whether directly, as in the case of plants via photosynthesis, or through the ingestion of plants by other organisms. The variation in natural light throughout the day and night is also of fundamental importance to life. The diurnal variation of light through the night and day acts as a *zeitgeber* or time giver that entrains the circadian rhythms of almost all organisms. Over the 3.8 billion years since life emerged, variations in light have come to choreograph many biological functions, from "sleep/wake cycles, sexual behaviour and reproduction, thermoregulation, and metabolic control such as energy intake/expenditure, glucose metabolism, lipid metabolism, and food and water intake" (Foster and Kreitzman, 2017, p. 36). Seasonal variations are also crucial in determining the timing of lifecycle events—when buds open, when animals reproduce and migration or hibernation patterns. These phenological cycles are driven by the changes in the diurnal cycle of light and dark throughout the year and the corresponding changes in temperature. These cycles and rhythms are all directly driven by the rotation of the earth on its axis and its orbit around the sun.

For millions of years, humans, like other life forms, have been bound to these rotational and orbital cycles. In recent centuries humans have developed technologies that increasingly allow us to subvert these biological imperatives and inhabit the night, if we so wish, as if we were nocturnal creatures. We are nonetheless diurnal; indeed, our eyes do not have *tapetum lucidum*—the crystalline retroreflectors that greatly improve the night vision of many nocturnal animals. To inhabit the night, humans rely on making the night brighter. As we have become better at doing this, we have changed the qualities of the night at an environmental scale, affecting not only our own circadian rhythms and behaviours but also those of other organisms and ecologies. This is increasingly recognised as a serious issue of many dimensions, from harming human health and wellbeing and disrupting ecologies to a loss of the night for terrestrial astronomy. It has also led to the loss of the night sky from social and cultural perspectives, eroding the sense of wonder and perspective that comes from the experience of being beneath a truly dark sky. Nonetheless, the perception of night-time lighting is strongly linked to human perceptions of safety and security on one hand and societal and cultural vitality on the other. Darkness, meanwhile, is often maligned as risky or dangerous, a place that invites accident, misfortune, and nefarious activity. This view of darkness fails to recognise the positive value of darkness in terms not only of its biological necessity for humans and non-humans but also its social and cultural importance—as a site for imaginaries that shift humans from centre stage and situate them in a more-than-human world, from the vastness of the universe in which we travel to the myths and mythologies of spirits and beasts. Such shifts in perception are pertinent as the effects of human exceptionalism become increasingly apparent through environmental degradation and climate change.

Unattended sensing

Sensing the Luminous Night was a pilot project that developed new methods of capturing variations in light at night, both natural and artificial. The research developed and installed a range of unattended light sensors in Leighton Moss RSPB Nature Reserve and developed a timepiece to communicate the data captured in a meaningful way. The sensors make periodic observations, capturing diurnal, circalunar, and annual rhythms of natural and artificial light, while the timepiece makes these temporal rhythms legible. Together, the sensors and timepiece were developed to augment our perceptions of darkness, facilitate social and cultural engagement with the night, and promote awareness, education, and discourse regarding the value of darkness.

On a practical level, the project installed long-range wide-area network (LoRaWAN) infrastructure for the sensors and an all-sky astronomical camera. It prototyped sensors that capture and transmit various parameters related to light during the day and night (lux, hue, 10-channel light spectra, ultraviolet and infrared light, temperature, humidity, and barometric pressure) and developed timepieces that visualise sensor data in terms of time and rhythm (Griffiths, 2023). In sympathy with the nature reserve's strong connection with birds, the sensors were housed in simulacra of eggs and fruit. A timepiece was chosen to communicate the data because clocks and watches are intuitively understood from both an embodied perspective and as an abstraction of times and timescales that we do not directly experience. Such a time-based approach can powerfully engage individuals with the natural cycles and rhythms of light and dark that we are familiar with and those of anthropogenic activity in our environments.

The timepiece shows time according to each of the various sensors, which capture observations every few minutes. The timepiece displays each reading as a dab of light on a screen (Figure 6.1), with the present moment always at the top of the screen and past readings slowly spiralling inwards. Each turn of the spiral represents a single rotation of the earth on its axis. As time goes on, readings accumulate to show the changing length of night and day and the variations in light during the day and night, such as daytime UV, seasonal changes in vegetation, the phases of the moon, and skyglow from nearby sources of anthropogenic light. Future work will install parallel sensors in a dark sky reserve to illustrate how much lighter the night has become due to human activity, even in a relatively dark place, such as Leighton Moss RSPB.

Through the various sensors, the timepiece brings several timescales into view: the momentary experience of environmental light as the sensor makes an observation; diurnal changes in light, both natural and artificial; circalunar variations in light, as the moon waxes and wanes; the annual changes in the length of day and night; and the phenological changes in the landscape over the year. The sensors were designed to capture observations over long periods and given time should show changes that occur over years, such as the year-on-year increase in intensity of artificial light (Kyba *et al.*, 2023).

Meanwhile, the all-sky astronomical camera takes a long-exposure photograph every 30 seconds between sunset and sunrise. The camera points directly up to the

Figure 6.1 Timepiece showing 24-hour changes in ambient environmental light over a period of approximately 30 days.

sky with a 180-degree field of view, creating a circular image framed by the silhouetted horizon. On a clear night, it captures the Milky Way or the movement of the moon across the sky. It also captures the sky glow from various urban settlements and the nuclear power station in Heysham to the west as well as more nuanced details, such as trains, cars, aeroplanes, and occasional meteors moving through its field of vision. Each morning, software automatically collates all the images that were captured into a timelapse video. This might show the night sky wheeling above and the regular passage of the local train service. Sometimes, it will show brisk clouds rushing across the sky, a bird's claw as it alights atop the dome, or raindrops collecting on the camera's transparent dome, small lenses themselves that distort the light from the sky through their luminous bodies.

The all-sky camera creates a detailed record of the night, which can be cross referenced to the timepiece. This can help to identify or confirm the source of particular changes in light and dark. While the clock gives a view of the night as a gradually changing continuum that choreographs behaviours and biologies, the all-sky camera gives a lively view of the night, full of weather, movement, and continual change. Although the pace is an illusion—the timelapse compresses the whole night into several minutes—the camera does give a particularly engaging view of the passage of the night and the cosmological, meteorological, and anthropogenic processes that animate it.

Together, these two forms of unattended sensing—the light sensors and the all-sky camera—provide a view of the night that emphasises duration, whether that be the short duration of a single reading or exposure or the much longer durations of diurnal and seasonal rhythms. Bringing such timescales into view creates temporal imaginaries within which we can locate direct experience.

Photography

The unattended sensors described earlier are used to create images and imaginaries that carry the changing patterns and rhythms of light and darkness over time. How do these differ from those created by the embodied experience of walking a landscape at night? Before answering this, we will first briefly consider a space between immobile sensors and the unconstrained mobility of walking, asking how the use of *attended* light sensors—digital cameras—might be used to experience, describe, and imagine dark places. Here, we focus specifically on a collaborative photographic practice in Leighton Moss RSPB (Figure 6.2). We approached this as experimental, playful, and open-ended, exploring the convergence of conviviality, technics, imagination, and a dark landscape. We will discuss just a few of the many observations that we made, hoping to nonetheless illustrate some key points regarding the relationship between photographers and the night that a camera helps to construct.

The nature reserve is very dark at night, and one must navigate via the sound of the ground underfoot, slight differences in contrast and the silhouettes of trees that give some slight indication of clearing or footpath. This encouraged us to settle in just a few locations—mainly a tower and a hide—rather than constantly roaming. These two sites suggested various image-making approaches. On a clear night, the tower—a three-storey structure open to the sky—offers views of the night sky and the hills that frame its edges, silhouetted by various shades of skyglow. Here, the camera became a collaborative form of surveying, searching for light and trying to identify its source. The camera was a prop through which to inhabit a space of

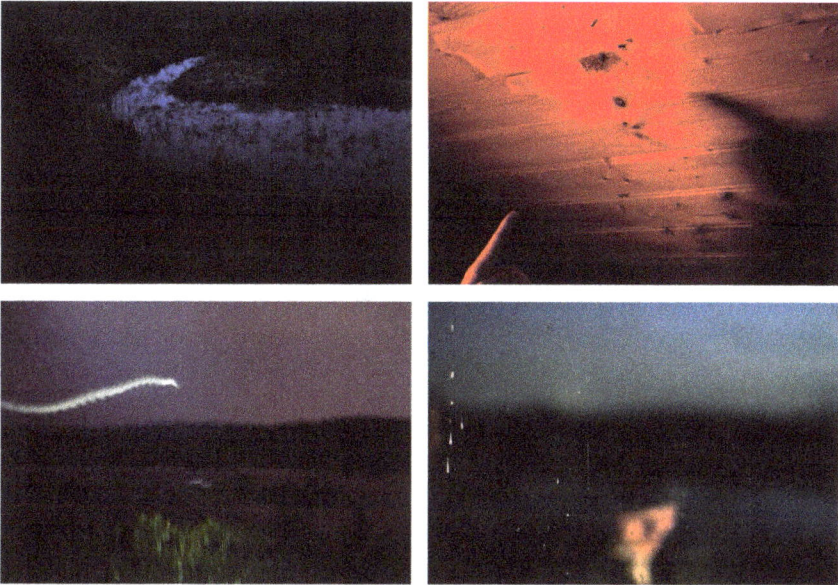

Figure 6.2 The landscape at nightfall—capturing the transient period between day and night. Clockwise: as the day fades into darkness, the water in the reedbeds comes to light; locating spider habitats in the hide; our image reflected in the hide window over the landscape; an insect's trajectory illuminated by flashlight during a long-exposure capture.

observation, not only through the lens but through shared presence and talking about what can (and cannot) be seen. We generally chose not to use tripods, preferring to be fully aware of the body's presence, capturing this through blur and shake or else using the affordances of the body and the space itself to prop the camera. The act of photographing becomes a collaboration with the distant landscape, the immediate site, and the body, as one braces and holds one's breath while pressing the shutter.

One visit saw heavy rain, and we went straight to the hide. Here, we had a view out over reedbeds occupied by various waterfowl. We began to capture long-exposure images of the landscape before us, casting a powerful flashlight in the heavy rain and capturing the trajectories of raindrops, tracing arcs as they struck the horizontal surface of lowered viewing windows. These images captured illuminated raindrops and the flicker of the LED flashlight, transforming their traces into a Morse code of light and dark. Switching to a red LED, we began to illuminate the immediate landscape, transforming the murky silhouette of a nearby sycamore tree into a preternaturally glowing red maple. As our torches fought the muddied light beyond the window, the resonant other worldly call of a stag drifted from the gloom into this briefly luminous space. As the night went on, we stopped and chatted. Outside, a row of birds settled into the night on wooden beams protruding from the water. Occasionally, they would rouse into a cacophony of noise, as if bickering about who should sleep where, and then just as quickly settle once more into silence. This went on for perhaps 20 minutes, and they were oblivious to our presence, despite the occasional sweep of red or white light from the torch.

These short examples illustrate how the practice of photographing positioned us in the environment and transformed it, sometimes using light (the flashlight), sometimes through sound (listening to the landscape and talking). The act of photography is a very specific way of being in the night, perhaps best described as an immersion. It requires and indeed creates full presence, sustained attention and a physical engagement with place and its agents. The photographic images, the temporal act of creating these images and the imaginaries that emerge as we do so inform and contribute to a thick description of the night environment, emphasising its potential for fictions and the imaginary. The camera itself both constrains and choreographs embodied rhythms of movement and stillness. These trajectories start and end with the dark pause of the shutter, an instant that presses the photographer's visual imagination against the landscape, leaving an indelible mark of one upon the other—the photograph itself. This slowing down of pace to capture an image allows the photographer to soak in new information and look for the particularities and idiosyncrasies of a place and the small details that animate images. The act of photographing was not taken as a means of creating a loyal representation of vision (our eyes are infinitely more sensitive) or a representation of an objective experience. Rather it was taken as an act of fictionalising the landscape through both intentional composition and fortuitous or accidental encounters. This act was one that nonetheless created heightened awareness and an intimacy with the shades of darkness that give substance to the landscape at night.

Walking

Night has been discussed as a place in its own right, fostering creative engagements and interpretations (Briggs, 2013; Bronfen, 2013). Although much has been written about walking as an (auto)ethnographic method for excavating the relationship between the (human) subject and place or landscape (Wylie, 2005; O'Neill and Roberts, 2020), little has been written about the walking in the night or walking with a view to understanding how the changing light of night and day affects our experience of place and landscape (Dunn, 2016; Edensor and Hughes, 2021) What perspectives and encounters does walking offer to an understanding of the night as a particular place and landscape? Here, we provide an example that illustrates how this question might be approached. It is an autoethnographic account of a nightwalk around the Leighton Moss RSPB Nature Reserve landscape. Walking at night enables the nocturnal sensible to be apprehended. It provides an embodied way to experience how different coexistences of light and dark constitute the identity of a place. Such an identity is dynamic, open, and provisional. Therefore, nightwalking contributes to the ways we might rethink how to conduct sensory ethnography (Pink, 2015). While there is a nascent strand of research on walking as a methodology through which to explore the more-than-human world, to date this has only been regarding the daytime (Springgay and Truman, 2018). By applying this approach after dark, it is our intention to demonstrate how walking can contribute to a thick description of the night and its imaginaries. Further, it enables us to also understand what areas could be augmented with additional methods.

A cold and dry night in late April. The sound of human feet along the boardwalk to the viewing platform offers a rhythmic dull thud announcing our presence to the non-human bodies fluttering and scurrying nearby. Behind us, the landscape seems to draw itself together, the darkness helping to gather its skirts up. Ahead of us, the moonlight stretches out, burnishing the occasional cloud with its silvering luminescence. It seems very quiet here. Standing still, we begin to tune into the soundscape. We listen carefully. We can hear the flora moving in the gloom, tendrils of dark green seem to flutter in the air around us. Whatever nyctinasty has been occurring around here with the onset of darkness, the message has not reached the wild garlic whose warm, pungent aroma drifts along to the left. Moving slowly towards the reed beds, unseen waterfowl stir and compose themselves. They ruffle about near the water's edge, aware of our company. As the path unfurls ahead, the viewing tower folds its way towards the sky, its stable and sharp geometry at odds with the undulating grasses and foliage around. Onwards and upwards. The horizon shifts with each step and turn until finally settling as we reach the platform and stop at its edge. We become profoundly aware of our bodies again, firm silhouettes against the night sky. From here, the dark mirror of the saltwater lagoon stretches across the landscape, smooth and crisp. Beyond, the hills tumble along and down into gloomy smudges. Above, the bruised sky shares its dark greys and murky blues and yellows. The wind seems to move through us, its soft caresses at ground level now firmer and indifferent to our bodies. The rushes offer their rustling applause as we descend the steps.

Threading between high grasses the boardwalk turns to gravel and compacted earth. We can sense water through the grasses and their gentle swaying invites our hands to move within their fibrous filigree. Far away, the call of an owl punctuates the night sky across the water. As we emerge from the parallel walls of reeds and grasses, the path sweeps into woodland. We enter its soft edges, and the landscape appears to wrap around us, a patchwork blanket of darknesses. Distance becomes harder to discern in here, the gloom rolls out in all directions, uneven. Now the ground yields to our boots as the tiny slide and stop of each forward motion navigates the mud. Tenebrous fingers of trees rush overhead to hold the night sky back. Fuzzes of bushes and undergrowth blur the edges of land and not. We become part of this sponginess, not entirely sure where we begin and the surroundings end, but instead there is an oscillation between the two. An interplay that stirs the senses and reshapes identities. The collective effervescence of this place in the daytime is replaced by something more subtle and sublime after dark. The air tastes of wood and wetland, its density almost perceptible in the mouth. The cushioned sounds of the landscape in the gloom are occasionally interrupted by the staccato flap or scuttle of our non-human neighbours in the nocturnal world. Focusing on what we can hear, smell, touch, and taste, we briefly feel assimilated into the landscape until the splosh of a puddle reminds us of our human awkwardness in this place. Following the slow rise of the path up to a road, the spell of the sensuous quietly retreats as we are confronted by the slick band of asphalt gliding away on either side of us.

Thick descriptions of the night

We have borrowed the term *thick description* from anthropology, where it refers to an ethnographic method for describing and interpreting culture (Ryle, 1971; Geertz, 1973). Unlike *thin descriptions*, which describe social observations, thick descriptions additionally describe the wider cultural context of such observations to interpret the meaning and value of actions and behaviours. Here, we use this term loosely, stretching and expanding it in several ways. First, we expand the context beyond human culture to include the environmental and planetary contexts within which human culture is held. Second, in addition to embodied field observation, we take the practice of observing to include mediated technological observations that extend our senses and imaginaries of darkness and the night, in this case via sensors and cameras. Third, in addition to texts and fieldnotes, we also record observations and make interpretations through data visualisations and photographs and reflections on the practices that create them. Finally, and perhaps most importantly, we take these practices—unattended sensing, photography, and nightwalking—to be not only the source of observations from which meaning and value can be drawn but also as sites where meaning and value are produced and enacted. By bringing these practices together, each attentive to our environments in different ways, the material and imaginary dimensions of the night and darkness are equally valued. Bringing these three practices together creates a site where embodied experience, memory, and imagination intersect with non-human and environmental scales of time and space—scales that sit outside of our direct embodied experience but that are nevertheless central to our biologies, behaviours, and societies and those of non-humans.

Capturing observations, whether by sensor, camera, or body, transforms landscape and subject in ways that combine notions of facsimile, imaginative interpretation, and fiction. At one end of our spectrum of night observations are sensors. These sensors capture data in a systematic, methodical, and consistent manner. Once installed, there is no human interaction with them—they simply do what they were programmed to do over and over. Of our methods, this is the closest to scientific observation. However, the approach taken to data visualisation takes the collection and communication of data as a social and cultural activity that, via the device of a clock, locates meaning in everyday life rather than in models of the environment drawn from data. It also provides a context not only for daily human activity but also the activities of all manner of birds, insects, plants, and animals. Photographic practice then intersects these shared environmental rhythms at a specific place and time—a couple of hours here and there in the night. Our presence immerses us wholly in these hours, obscuring the context of longer periods of time—the cyclical progression of days, months, and years. Nonetheless, the lineage from sophisticated professional cameras to the one-pixel cameras of the sensors connects us back to this context. A line is drawn between the wider environment and our fleeting presence via light-transforming technologies, whether camera or sensor. This presence also intersects perambulatory bodies with senses relatively unmediated as they move through landscapes (bar the important and transformative technologies of footwear and clothing). Thus, three trajectories are drawn: one from disembodied data collection to embodied observation, one from environmental rhythms to bodily rhythms and sensation, and one from apparent facsimile of the environment to imaginative interpretation and fiction. Together, these create a thick description of the night that is attentive to and interpreted though multiple temporalities, bodies, and technologies.

Conclusion

This chapter has presented three ways through which we can apprehend the sensitivities of places after dark. Such knowledge is crucial if we are to develop appropriate principles and practices for urban design and policy to support multispecies life during both the day and night. By elucidating on the ongoing entanglements between light and dark, bodies and landscape, and time and space, we have sought to demonstrate the need for thick descriptions to better understand dark places. While there is increasing care and attention devoted to the preservation of darkness in key places, as assiduously championed by DarkSky (2023), this has largely been in relation to rural sites that are relatively remote from urban environments and, by extension, the impacts of light pollution. Yet the conversation with regard designing with darkness in urban places remains nascent. Current practice by built environment professions and policymakers tends to draw upon long-held binary view of light and dark, with ALAN synonymous with safety and security despite successive implementations resulting in urban illumination that is often excessive if not downright careless. Through producing thick descriptions of places after dark to inform a framework for a more temporally sensitive approach to urban planning and policy (Gwiazdzinski, 2015), we can overcome the misunderstandings that are associated with darkness and urbanism. These thick descriptions help us to become more attentive to darkness and its value for humans and non-humans. We propose that such knowledge can inform built environment design practices and policymaking by enabling them to be heedful to the various ways in which coexistences of light and dark and of humans and non-humans shape the places we experience. As light pollution now presents a global challenge, acknowledging the diverse interplay that darkness and light provide is critical for the transitions necessary to tackle its impacts on humans and non-humans in a local and situated manner. For this to be effective and empower us to rethink places at night, especially urban environments, we need new nocturnal imaginaries (Dunn, 2023). Being able to account for the alternative knowledges of different relationships and values across dark places will enable the creation of nocturnal imaginaries that can shape the urban night and vice versa. To conclude, adopting these methods to inform how design can have greater nuance toward the coproduction of places at night would support a more-than-human approach to their ongoing development over time.

References

Briggs, B. (ed.) (2013) *3Am: Wonder, paranoia and the restless night*. Liverpool: Liverpool University Press and the Bluecoat.

Bronfen, E. (2013) *Night passages: Philosophy, literature, and film.* New York, NY: Columbia University Press.

DarkSky. (2023) *DarkSky.* Available at: http://darksky.org/ (Accessed: 1 August 2023).

Dunn, N. (2016) *Dark matters: A manifesto for the nocturnal city.* New York, NY: Zero Books.

Dunn, N. (2023) 'Nocturnal imaginaries: Rethinking and redesigning the city after dark', *Ethnologies,* 44(1), pp. 107–128. https://doi.org/10.7202/1096059ar.

Edensor, T. and Hughes, R. (2021) 'Moving through a dappled world: The aesthetics of shade and shadow in place', *Social & Cultural Geography,* 22(9), pp. 1307–11325. https://doi.org/10.1080/14649365.2019.1705994.

Foster, R. and Kreitzman, L. (2017) *Circadian rhythms: A very short introduction.* Oxford: Oxford University Press.

Geertz, C. (1973) *The interpretation of cultures* (Vol. 5043). New York, NY: Basic Books.

Griffiths, R. (2023) 'Time and the Anthropocene: Making more-than-human temporalities legible through environmental observations and creative methods', *Time & Society,* 0(0). https://doi.org/10.1177/0961463X231202928.

Gwiazdzinski, L. (2015) 'The urban night: A space time for innovation and sustainable development', *Articulo—Journal of Urban Research,* 11. https://doi.org/10.4000/articulo.3140.

Kyba, C. C. M., Altıntaş, Y. Ö., Walker, C. E. and Newhouse, M. (2023) 'Citizen scientists report global rapid reductions in the visibility of stars from 2011 to 2022', *Science,* 379(6629), pp. 265–268. https://doi.org/10.1126/science.abq7781.

O'Neill, M. and Roberts, B. (2020) *Walking methods: Research on the move.* New York: Routledge.

Pink, S. (2015) *Doing sensory ethnography.* London: Sage.

Ryle, G. (1971) 'The thinking of thoughts: What is "Le Penseur" doing?' in Ryle, G. (ed.) *Collected papers, Vol. II: Collected essays 1929–1968.* London: Hutchinson, pp. 481–496.

Springgay, S. and Truman, S. E. (2018) *Walking methodologies in a more-than-human world: WalkingLab.* London: Routledge.

Wylie, J. (2005) 'A single day's walking: Narrating self and landscape on the South West Coast Path', *Transactions of the Institute of British Geographers,* 30, pp. 234–247.

7   Considering festive illuminations in dark sky places

Honouring darkness, creative innovation, and place

Tim Edensor and Dan Oakley

Introduction

Across the world, an extraordinary range of light festivals continue to emerge as attractions and events that entice locals and tourists. Diverse cities recruit design professionals and artists to transform their nocturnal environs over times that stretch between a few days and several weeks, and an increasingly variegated array of technologies are deployed to beguile attendees. Large-scale, capital-intensive events lure tens of thousands, even millions, of visitors, such as Lyon's Fête des Lumières and Sydney's VIVID festival, and draw upon an international network of designers, managers and technicians (Giordano and Ong, 2017). There is also a plethora of smaller, local festivals, including lantern parades and processions (Edensor, 2018; Skelly and Edensor, 2020), as well as technologically enhanced traditional festivals, often sacred in origin, such as Diwali and Loi Krathong. The variety of lighting techniques, forms, installations, and fixtures includes light projections, laser displays, interactive works, street furniture and animation that cast light across space in multiple hues and tones. Such displays manifest the potential of illumination to enchant and defamiliarise places, generate convivial atmospheres and foster playful, interactive engagements.

In this book, chapters reveal that arts-led festivals are also progressively being curated in dark sky settings, and they disclose that these events are sidestepping the use of bright lights in innovative and creative ways. However, this is not so in the case of the South Downs International Dark Sky Reserve (or Moore's Reserve, named after renowned public astronomer, Patrick Moore), assigned in 2016, only 100 kilometres away from London and exceeding 1,500 square kilometres. It sits within the South Downs National Park. Residents and businesses within the park and in adjacent protected spaces continue to stage festivals of illumination, constituting a difficult challenge for managing authorities.

Given the immense spread of nocturnalisation and the colonisation of most spaces with illumination (Koslofsky, 2011), it may be regarded as inapposite to stage any illumination at a dark sky park. Given the pre-eminent goal of preserving darkness, as well as the awe solicited by an encounter with the colossal light display of a vibrantly starry sky and the immensity of the universe, any light-oriented events may seem gratuitous. Yet at South Downs DSR and in nearby

DOI: 10.4324/9781003408444-10
This chapter has been made available under a CC-BY-NC-ND 4.0 license

protected spaces, commercial and tourist concerns have prevailed over such conservationist ideals; accordingly, at present the authorities have decided that it is more practical to work with festive events and offer advice on how some of the most malign effects of lighting might be ameliorated. Several profit-led ticketed events, some with an emphasis on spectacular features, contrast with other more artist-led festivals, usually free-to-view, in which thought-provoking, subtle, arresting and place-specific installations tend to be privileged. Because of their limited operational time, events of this type are not usually subject to planning control and are considered as "temporary" developments. As such, their impact on the local environment is rarely considered (Oakley, 2023). In this context, the South Down National Park Authority has adopted a pragmatic approach that involves discussion with and the education of the organisers of the larger events, and this has involved the formulation of specific guidance notes. However, a problem faced by managers of dark sky places is that there is no manual, no delineated standards or internationally accepted definition to guide how lighting design should be deployed in such settings. Recourse to the interpretations provided by definitions of light pollution—trespass, skyglow, glare, and intrusion—as with the guidance offered by GN01 Obtrusive Light by the Institution of Lighting Professionals, can be measured by photometry. As such, guidance can include recommendations that festival organisers can minimise light pollution by adopting measures like "Avoiding cool white light, Reducing upward light, Turning off light at close of event, Using wider beam angles on more powerful spots, Lowering the power" (ibid., p. 6).

One quality that dark sky realms provide in foregrounding the potency of light is to enhance the aesthetic and artistic qualities of illumination by locating them within a gloomy setting and against a very dark background. In contrast to most urban settings in which multiple sources of light are deflected, reflected, and absorbed across space, generating an unpropitious environment in which striking forms of illumination lose their allure as they merge with the general ambient array, the full form, glow, intensity, and colour of a light installation may be far more effectively and dramatically discerned against a dark backcloth.

This chapter assesses the suitability of the limited management options currently available to national park and dark sky reserve managers in the South Downs to reduce the effects of over-illumination. We firstly highlight some of the most deleterious effects on darkness perpetrated by festive lighting before discussing creative, sustainable, and place-specific examples of illumination that more appositely mobilise more progressive approaches to engaging with gloom and place. We discuss the desirability of producing more influential guidance for light festival organisers and technicians, before exploring how both urban and rural examples point to the use of inventive but less intrusive illumination and conclude by suggesting that these examples might be extended beyond officially designated spaces of protection and conservation. We ask: how might light be used effectively to enhance place after dark, be sensitive to non-humans and complement the allure of darkness?

Inappropriate lighting in South Downs Dark Sky Reserve

Temporary car park floodlighting

At several events in South Downs DSR and other nearby dark landscapes, the greatest threat posed to the surrounding darkness does not consist of the light shed from any of the festive lighting installed. Perversely, the lights that have by far the biggest impact are not included as part of the initial design; they are temporary security mounts, typically those that require generators and extendable poles. Such lights are primarily used for car parking, security, and safety in areas in which people and cars share the same space or where crowd control is important. In dark, rural landscapes, these very intense, high-glare luminaires often excessively illuminate a dark area and create significant sky-glow domes above as the light scatters off surfaces. Moreover, the power of these lights is sufficiently high that it can create significant glare issues and safety issues for oncoming drivers. Set up badly, these lights can be a liability if they are used at their default maximum setting, creating dangerous contrast issues where pedestrians can be hidden behind blinding light. Indeed, this maximum intensity is most frequently mobilised without consideration for meeting the illuminance criteria recommended by the British Standards or HSE, with little thought given to setting up the right column height for the area or the appropriate colour temperature or even the upward trajectory of the beam. This indicates that this is not solely the fault of the of the makers of this lighting; rather, more guidance is needed to enable event organisers to use and install this lighting in situ. More judicious applications could be achieved through minor adjustments.

Sky scanners, searchlights, and spotlights

Of equal concern to event managers and light designers is the continued presence of sky scanners and searchlights, often using high-powered lasers that seem to create the antithesis of a dark sky aesthetic. Such illuminations can be seen from several miles away and send an intense, tightly angled beam of light high into the atmosphere that often dances along the bottom of cloud bases; on especially overcast nights, this can create significant sources of light pollution, causing clouds to brightly glow. While gratuitously used for marketing purposes on occasion, turning the sky into a billboard, searchlights, sky scanners are undoubtedly dramatic as navigational aids and are alluring to those on their way to a light festival. Yet they vanquish any sense of a dark sky and in their generic form rarely add to any particular sense of place. Better modes of way-marking can avoid illuminating the cloud canopy by lowering the angle of the beam and directing it away from dark areas. Nonetheless, it is preferable to remove all presence of searchlights and sky scanners from dark sky lighting festivals and seek more sustainable forms of effective signage.

Theatrical spotlights have also become a common addition to the festive illumination inventory in contributing to a dynamic and colourful display. Over time, they have become more abundant, becoming cheaper, more waterproof, and easier to control by being plugged in and synchronised with predetermined movements. Besides directing audiences to precise locations, such lights may be used to communicate a narrative in the transmission of dramatic stories. While these lights can be obtrusive, particularly if high power and tight beam angles are used, many have

Figure 7.1 Illuminated tree surrounded by festoon lighting, South Downs National Park. Photograph by Dan Oakley.

an in-built capacity to change their beam width, angle, and colour, retaining an intimacy within a particular space. By widening the beam angle, the cross-sectional area of light can be diluted and penetration into the night sky reduced, an effect that can also be achieved by reducing the upward angle. Moreover, redder light temperatures can minimise the penetration in comparison to white light and reduce the impact on visual adaptation to the dark.

As at other rural light displays, certain spotlights in the South Downs DSR and adjacent spaces have sought to focus on individual trees to delineate their form against a dark backdrop (Figure 7.1). Such shapely prominence could not be achieved during the ambient conditions of everyday daylight in which objects can be less easily sepa-rated from each other, foregrounding the capacity of illumination to light up a discrete entity that distinguish it from its surroundings and thereby offer an alternative mode of perceiving, allowing a visual focus on form and texture that fosters appreciation of the separate constituents of the world. Yet in undertaking such techniques, certain measures should be adopted to minimise the disturbance to non-humans, notably during the bird-nesting season. Low-powered luminaires, ideally with a lumen output restricted to 500, should be deployed, and light should be absorbed by canopies at as low a point as possible, avoiding the lighting of tall trees with sparse foliage.

Unless modest in colour temperature and intensity, spotlights may contribute to skyglow. This has been largely avoided in the case of an astute application of spot-lights at a National Trust site in a protected landscape adjacent to the South Downs National Park. Designed to communicate ideas about the critical role humans must play in stewardship of the Earth, numerous spotlights were arranged to coordinate with music so that they could dance across the woodland textures at the edge of a small lake. At key points, spotlights directed the visitor's gaze towards illumi-nated models of planetary objects installed around the water as the music fluctuated between major and minor strains. The display reached a crescendo as the lights were directed towards an illuminated Earth hanging in space. The message was as simple as it was obvious: protect the Earth.

Digital mapping and projection

Using building facades to illuminate architectural features is a tried and tested method to theatrically enliven spaces, with prominent buildings highlighted to form spatial focal points or to guide the eye towards important landmarks. At the sim-plest level, fixed colour washes are used to apply a background colour and accen-tuate building features. A more dynamic and enormously popular technique used by designers and artists at light festivals is the use of building facades as screens upon which digitally mapped images are projected. Digital mapping techniques deploy software to exactly "map out points of interface with the spatial coordinates of the existing structure" using specific architectural features—windows, niches, cornices, columns, and arches (Susik, 2012, p. 115). Site-specific projections can startlingly alter the usual apprehension of the building, transforming colour and animating textures and forms, sometimes producing illusory effects that suggest that buildings are dematerialised or composed out of unlikely materials.

Abigail Susik (2012) argues that the use of digital light projection by commercial interests could saturate illuminated space with a thoroughly mediatised architecture, replete with a blizzard of images and messages from multiple angles and planes. However, at the Cowdray estate at Midhurst within the South Downs National Park, the great Tudor house ruins were used as a backdrop for a colourful and dynamic display that told of the estate's history from house to garrison and destruction (Figure 7.2). In folding together different times and spaces, the projection focused on how architectural designs and manual toil were integral to the ongoing transformation of the house over the centuries. Besides deepening historical understanding and architectural awareness of key features, the projection also conjured up a sensory feel for the past, with sounds of labour and hard physical endeavour honouring forgotten labourers and artisans who contributed to the ongoing emergence of the building. This illustrates how pessimistic future scenarios can be countered with the alternative that such techniques offer abundant promise for creativity and dynamism. These effects are also shaped by how digital mapping is deployed, with inappropriate projections using excessively bright lighting or misdirected light so that light trespass strays beyond the façade and contributes to skyglow, a point of contention at Cowdray.

Figure 7.2 Projection mapping on the Cowdray estate, Midhurst. Photograph by Dan Oakley.

Place-based, modest, and creative illumination in South Downs Dark Sky Reserve

In addition to these widespread spotlighting and projection applications, light festivals have also included a medley of smaller, less intrusive, creative forms of illumination, installations of increasing diversity in form, size, colour, and intensity. Especially notable are those bespoke creations devised for a particular event and setting, underlining the salience of place-specificity while also providing opportunities to defamiliarise familiar space through which nocturnal space is experienced as suddenly strange (Edensor and Sumartojo, 2018). Where modest in size, such installations may provide visitors with more intimate experiences than projections and larger pieces from which they must keep a spectatorial distance. Even more enchanting are those works that offer interactive and playful engagements that, for instance, may respond to movement or sound, reflect back images and shadows of onlookers, or solicit movement across their illuminated forms.

Other lighting applications are especially well suited to dark sky environments. Festoon lighting is particularly unobtrusive. Here, an assemblage of many low-powered lights, such as strings, lanterns, fairy lights, or bulbs, are used to light spaces and paths (see Figure 7.1). As long as individual lights retain sufficient distance from each other to avoid bunching, the overall effect can be aesthetically subtle and avoid the dispensing of large quantities of intense lighting to dark spaces. Another increasingly popular technique is to create structures out of canvas, plastic, or glass that are subsequently lit from within by low-powered illumination to produce a glow that sets a structure apart from the surrounding environment, but which distributes little superfluous lighting upon surrounding space (Figure 7.3). Against dark sky backgrounds, such glowing installations can be especially effective, their illuminated shapes standing out starkly whilst minimising light pollution, luring onlookers from a distance to discern their forms and adding qualities of mystery and enigma. Such lighting underpins how illumination does not need to be especially bright in order to make an impact and how deep darkness can enhance its potency. Such artistic pieces can transform places into realms of surprise, fascination, and sensory stimulation. In developing a focus on these more place-specific and discreet installations, we now identify four particular installations in the South Downs DSR that foreground how subtle lighting can contribute to memorable and richly sensual experiences.

First, at the Wiston Estate in the South Downs DSR, light works are annually installed in winter across the gardens, buildings, and parkland conjoined by a nocturnal promenade (Light Up Trails, n.d.). The event has been subject to some criticism for vanquishing the gloom that the designated dark sky realm is intended to provide, with certain bright illuminated displays obliterating darkness. However, other, subtler illuminated displays combine with place-specific materialities and textures, as well as the surrounding darkness, to produce impactful experiences. One such work is *Harmonic Portal*, by Chris Plant, an installation that use saturated, ever-changing colours inside and outside of a circular frame to intensify a sense of the lithic and vegetal textures of local surfaces, vividly disclosing qualities that may not ordinarily be noticed (Figure 7.4). At Durham Lumiere in 2017, the

Figure 7.3 Illuminated canvas installation, South Downs National Park. Photograph by Dan Oakley.

Figure 7.4 Chris Plant, *Harmonic Portal*. Photograph by Tim Edensor.

work was placed against a medieval stone wall; at Wiston, a series of installations were situated on an ivy-clad wall, a stone wall and a thick hedge, intensifying awareness of the particular material textures and aesthetics of place.

Second, at a Kew Gardens property in the High Weald Area of Outstanding Natural Beauty (AONB) in Surrey, a low-lying small pond surrounded by trees was transformed through the ingenious deployment of luminaires. The overhanging trees clothed the area in gloom that largely shielded visitors from seeing the site's other illuminations. As the path descended to the pond, visitors were greeted by small installations resembling blue creatures that guided the way into a space that was washed with glowing, low intensity, blue and purple LED floodlights. In the overhanging trees, more blue creatures pulsed slowly, bestowing upon the entire space a dynamic, animated quality (Figure 7.5). Yet while looking from a distance, this illuminated space was barely discernible. The light designers had cleverly used the configurations, contours, and vegetal growth of the site to situate lights that dramatically altered the look and feel of the place while ensuring that this neither intruded into other realms of the estate nor impaired the qualities of the dark skies. As with the planetary installation referred to earlier, which was situated on the same estate, a reflective space was curated rather than simply a collection of illuminated adornments.

Third, an example of a rather different innovative lighting installation devised to mark a sense of place are the illuminated sheep sited at the Lewes Light Festival in

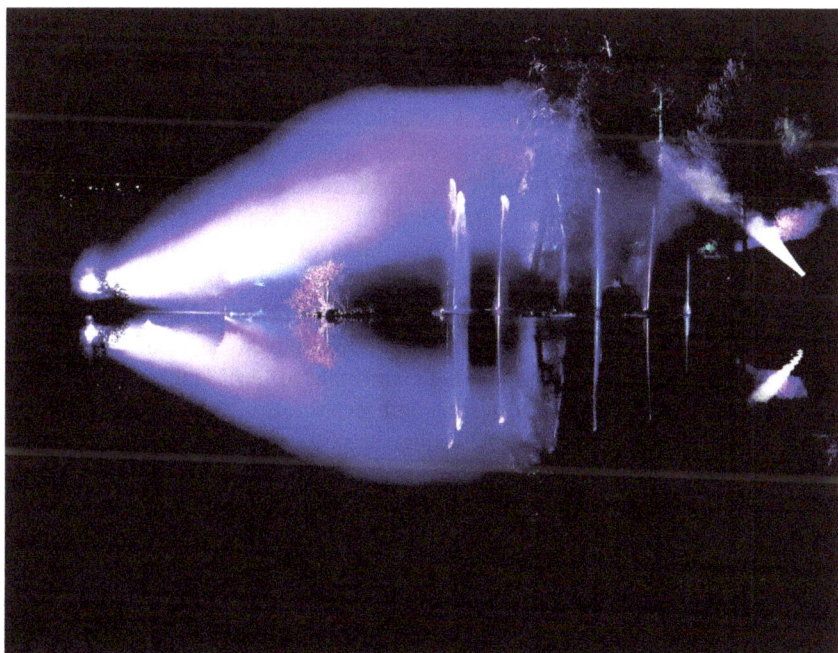

Figure 7.5 Kew Gardens, illuminated pond and surroundings. Photograph by Dan Oakley.

Figure 7.6 South Downs sheep, Lewes Light Festival. Photograph courtesy of James McCauley.

2018, an event that is regularly staged in the winter months. This is predominately a free urban festival where local designers and groups work together to produce artistic installations that complement the cultural heritage of the market town. In this spirit, project manager Graham Festenstein commissioned a series of iconic life-size, internally illuminated South Downs sheep (Figure 7.6). Lewes, located within the South Downs National Park, is contextualised by its rural surroundings, specifically the rolling chalk grassland hills that dominate much of the views from the high street and castle. This rare habitat that lies atop a calcareous soil has been historically maintained by grazing animals, primarily sheep, which keep the sward low, increasing the biodiversity of the habitat. Much of the South Downs and Lewes's historic economy was based upon the effectiveness of this grazing, and sheep thus retain a highly symbolic connection to local landscapes and their management. The illuminated sheep remind locals and visitors of the important role of Lewes as a market town and confirm the historical link between the town and the downs.

A final example from the Lewes Light festival signifies how artificial illumination might be complemented by a focus upon the dark skies visible from a nearby hilltop on which the South Downs National Park Authority held a "telescope star party". Halfway along the lengthy walk from town to the fairly dark location, a glow worm installation was created, formed of low-powered single green LED lights nestled within a tree that mimicked this iconic regional insect species in larval form. The glow worm provided a visual reference that connected the town to its local rural nocturnal landscape while providing a bridge to the star-filled skies visible from the telescopes.

Reconsidering festive and creative design in dark sky places

It would be ideal if light festivals ceased to be staged in the South Downs DSR and other protected landscapes but given their popularity and commercial value, this outcome seems unlikely to occur in the near future. It is problematic that temporary lighting events in otherwise protected landscapes do not usually require planning permission and are not subject to any consultation by the local planning authority. Accordingly, the National Parks Authority must accept such intrusions and hope that the designers will work with them to minimise malign effects. While we have noted the guidance provided by Dan for such events in the local setting of the South Downs National Park, it would be preferable if more extensive advice were available that more authoritatively circumscribed what kinds of lighting were permitted at such events. It remains anomalous that DarkSky delineates its "Five Points for Responsible Lighting", yet unregulated illumination may be temporarily installed in a wide range of dark places. National or international guidance that delimit the deployment of lasers, floodlights, searchlights, excessive way-marking and car-parking luminaires, and over-radiant digital mapping in such spaces would help to minimise the baleful effects of over-illumination and protect the night sky.

More progressively, given the expanding advances in lighting technologies and artistic uses, a substantive catalogue of possibilities might also assist festival managers in choosing lighting that registers the qualities of place and landscape without presenting a discordant intrusion into an otherwise dark place that negatively affects non-humans. Indeed, it is evident that the judicious deployment of low-level, artistic, and site-specific illumination that more minimally affect dark sky settings can be extremely effective in charging dark realms with rich atmospheres and defamiliarising effects, reconfiguring sensations and deepening a sense of place, besides enhancing social and environmental sustainability. Here, we have identified festoon lighting, illumination installed from within opaque structures, and lighting that aligns with contours and tree cover to ensure that it is carefully concealed from a distance and shines only within discrete places. We have also exemplified how low-level lighting displays can enhance a grasp of the textures, histories, and environmental contexts of place in subtle and revealing ways. Moreover, such lighting may lure visitors so that festive events might serve as appropriate stepping-stones to familiarise visitors with designated dark sky settings and thus extend knowledge and understanding about their purpose and worth, connecting people to the nocturnal sky and bring them on board as champions and advocates.

It is timely to consider that the commercial attractions of darkness for tourists and urbanites, especially when they are ameliorated by minimal forms of illumination, are not new.

The potential for the development of diverse experiences of gloom is also exemplified by the range of dark experiences that have been developed in primarily urban interior places over the past few years to satisfy increasing desires to encounter darkness and escape from the inexorable presence of bright lights. These include the simulacra of famous urban sites that have been devised for tourists to convey sensory experiences beyond the visual through dark exploration. Similarly, dark dining has emerged to focus on the sonic, tactile, social, aromatic, and tasteful qualities of eating out while concerts staged in the dark attune audiences to a heightened appreciation of music without visual distractions (Edensor and Falconer, 2014). In addition, diverse theatrical events and participatory engagements (Edensor, 2017; Welton, 2020; Sumartojo, 2020) imaginatively enjoin attendees to become immersed in imaginary and dramatic engagements. Although such enticements are yet to be developed on any substantial scale in outdoor places and landscapes, there seems little reason that they could not be transposed to an array of potent dark sky settings.

In a plethora of urban settings, the manipulation of light, shadow, and shade has been integral to both historical, modern, and contemporary architectural practices as Stephen Kite (2017) extensively details. Once more, such deployments could also influence the creation of rural buildings. Similarly, the creative insights of a group of some 400 light designers who call for the more sustained deployment of what they term the "dark art" could inform the landscaping of rural dark sky places in ways that enhance the experience of darkness. In reclaiming gloom as a positive quality, they declare "Let there be darkness", promoting aesthetic qualities of mystery, imagination, chiaroscuro, theatricality, and atmospheric potency. Chris Lowe and Philip Rafael (2020), members of this group, discuss how divergent elements of the nocturnal landscape can be highlighted to foster a greater sensitivity to colour variation, shape and surface, and modest luminaires can reveal how light is refracted, absorbed, or reflected by the materialities with which it interacts. They also call for a greater focus on how light and dark can subtly mark the passing of circadian and seasonal rhythms and be organised to engender choreographies through which people negotiate space. Such modes of moving through dark rural space are evocatively discussed by Ellen Jeffrey in this volume, and the more conscious design of rural dark spaces is also echoed in Gernot Böhme's (1993) observation about how light, shade, and shadow can be orchestrated in landscape gardening as practices that "tune" space by managing the levels of light that filter through woodland canopies.

Besides these imaginative encounters with darkness, other inventive recent theatrical attractions that take place under rural dark skies have depended upon the potency of darkness for their efficacy. Nina Morris's (2011) account of *The Storr: Unfolding Landscape*, a dramatic 2005 event created by the Scottish arts group, NVA, which ran across a spectacular mountainside on the Isle of Skye, describes how visitors moved across hillsides and through woods after nightfall, experiencing a rural darkness that was periodically interspersed with diverse light projections, mini-dramas, poetry, and music. As a richly "textured realm of sensory perception" (Morris, 2011, p. 335), darkness acted to reconfigure apprehension of the landscape and set the scene for the theatrical interludes.

Finally, creative installations, events, and techniques might be adopted to solicit wider awareness of these diverse gloomy qualities. One such fixture devised to encourage reflexive, meditative attunements to shifting conditions of light and dark in the landscape is detailed by Nina Morris (2015) in her depiction of Dalziel and Scullion's 2013 non-luminous, hard plastic benches positioned in a clearing in Galloway's dark sky park. Set amidst the coniferous forest and fashioned with a shallow depression at one end, visitors are encouraged to lie on their backs, resting their heads as they look upwards to the night sky. The designers describe the benches as an "invitation and a conduit to the sensorial resources of hillsides, woodlands, loch sides and riverbanks" (Dalziel and Scullion, n.d.).

The visitor experiences solicited by these events and installations of the shift through the stages of twilight towards nightfall conjure up Peter Davidson's (2015) extraordinarily subtle account of the ever-changing, sensory, and metaphorical potency of civil, nautical, and astronomical twilight, conditions that he personally experiences and cites their representation in art and literature. His rich, poetic writing about the shifting hues and shadowy forms that fade and emerge over the diurnal cycle exemplify how programmes to accustom visitors to the different qualities of gloom that devolve in dark sky places might be more substantively developed, moving away from artificial illumination and countering an era in which such transitions are overlooked. Similarly, the sky spaces created by James Turrell provide an intense focus on the gradual changes from daylight to the night sky that occur through the stages of twilight and over the seasons (Edensor, 2017). In these rural settings, such transitions from light to dark may be experienced without the intervening distractions of artificial illumination, once more underlining how places designated as dark sky realms offer enormous potential for experiencing the multiplicities of darkness.

References

Böhme, G. (1993) 'Atmosphere as the fundamental concept of a new aesthetics', *Thesis Eleven*, 36(1), pp. 113–126.

Dalziel and Scullion (n.d.) *Rosnes bench*. Available at: https://dalzielscullion.com/works-entry/rosnes-bench/. (Accessed: 10 January 2023).

Davidson, P. (2015) *The last of the light: About twilight*. London: Reaktion.

Edensor, T. (2017) *From light to dark: Daylight, illumination and gloom*. Minneapolis, MN: Minnesota University Press.

Edensor, T. (2018) 'Moonraking: Making things, place and event', in Price, L. and Hawkins, H. (eds.) *Geographies of making/making geographies: Embodiment, matter and practice*. London: Routledge, pp. 60–75.

Edensor, T. and Falconer, E. (2014) 'Eating in the dark: Sensation and conviviality in a lightless place Dans Le Noir?' *Cultural Geographies*, 22(4), pp. 601–618.

Edensor, T. and Sumartojo, S. (2018) 'Reconfiguring familiar worlds with light projection: The Gertrude Street Projection Festival, 2017', *GeoHumanities*, 4(1), pp. 112–131.

Giordano, E. and Ong, C-E. (2017) 'Light festivals, policy mobilities and urban tourism', *Tourism Geographies*, 19(5), pp. 699–716.

Kite, S. (2017) *Shadow-makers: A cultural history of shadows in architecture*. London: Bloomsbury.

Koslofsky, C. (2011) *Evening's empire: A history of night in early modern Europe*. Cambridge: Cambridge University Press.

Light Up Trails. (n.d.) Available at: https://www.lightuptrails.com (Accessed: 8 January 2023).

Lowe, C. and Rafael, P. (2020) 'Designing with light and darkness', in Dunn, N. and Edensor, T. (eds.) *Rethinking darkness: Cultures, histories, practices*. London: Routledge, pp. 216–228.

Morris, N. (2011) 'Night walking: Darkness and sensory perception in a night-time landscape installation', *Cultural Geographies*, 18(3), pp. 315–342.

Morris, N. (2015) 'Exhibition review: Dalziel and scullion, Tumadh: Immersion', *The Senses and Society*, 10(2), pp. 261–266.

Oakley, D. (2023) *South downs national park: Design guidance for creative events, design commissions and lighting festivals in dark skies*. https://darksky.app.box.com/s/ii8gm9jnuasf97f24744trxw17l0ju82/file/1124964862674.

Skelly, G. and Edensor, T. (2020) '*Routing* out place identity through the vernacular production practices of a community light festival', in Courage, C. (ed.) *The Routledge handbook of placemaking*. London: Routledge, pp. 258–268.

Sumartojo, S. (2020) 'On darkness, duration and possibility', in Dunn, N. and Edensor, T. (eds.) *Rethinking darkness: Cultures, histories, practices*. London: Routledge, pp. 192–201.

Susik, A. (2012) 'The screen politics of architectural light projection', *Public: Art, Culture, Ideas*, 45(1), pp. 106–119.

Welton, M. (2020) 'Going dark: The theatrical legacy of Battersea arts Centre's playing in the dark season', in Dunn, N. and Edensor, T. (eds.) *Rethinking darkness: Cultures, histories, practices*. London: Routledge, pp. 179–191.

Part 4

Non-human entanglements with dark skies

8   Nature's calendar, clock, and compass

What happens when it is disrupted

Therésa Jones and Marty Lockett

Introduction

In this chapter, we explore the role of light (sun, moon, stars, and artificial light at night [ALAN]) in the evolution and maintenance of life on earth. We first outline how natural light has shaped seasonal, monthly, and daily cycles. We then highlight how the increasing presence of ALAN has disrupted these natural cycles and how different aspects of light influence behaviours, physiologies, and ecosystems. Finally, we explore the conflict between human desires and needs for nocturnal brightness, its impact for the rest of life on the planet, and offer suggestions for mitigation and resolution that will benefit all species, including humans.

Natural light cycles

Life on Earth has evolved over the past four billion years under relatively stable light cycles. Sunlight and moonlight wax and wane in patterns that result in daily (day-night), monthly (lunar), and annual cycles (shifts in the duration of daylight dictate the changing of the seasons). The consistency of these cycles is such that a tyrannosaur travelling forward 70 million years would find today's day-length changed by less than half an hour (de Winter *et al.*, 2020). Such predictability also means that until recently, for a given location on the planet, the amount of sun- or moonlight visible on any day of any month in any year can be estimated to a fraction of a second. These two factors—predictability and consistency—mean that natural patterns of light have always been more reliable biological indicators of time compared to other abiotic signals, such as temperature or humidity. Even in the face of the recent dramatic changes in climate that are disrupting biological systems globally, natural light cycles remain relatively consistent.

Of course, natural light cycles do not run independently of other abiotic factors. Sunrise is associated with the onset of increased light and warmth over a 24-hour period, the more visible, fuller moon aligns with brighter nights and stronger tides, and the shortening length of daytime signals a cooler and darker period of the year. Nonetheless, light most reliably foretells changing conditions, and biological systems predominantly use light as their daily clock and yearly calendar. Accordingly, nearly all facets of biology, from behaviours through to physiologies, have

DOI: 10.4324/9781003408444-12
This chapter has been made available under a CC-BY-NC-ND 4.0 license

evolved in response to, and are intricately governed by, natural light cycles. Light facilitates activity, growth, development, reproduction, migration, and hibernation of organisms at optimal times of the day, month, or year. Animals temporally partition themselves across a 24-hour day, resulting in diurnal versus crepuscular versus nocturnal species (Tierney *et al.*, 2017). They variously sing and hunt more (or less) often under the light of a full moon (Mougeot and Bretagnolle, 2000; Dickerson, Hall and Jones, 2020) and reproduce during times of the year when resources are plentiful and predation threats are fewer (Robert *et al.*, 2015). Similarly, plants photosynthesise during the day, and changes in day-length signal the onset of periods most amenable for germination, new growth, flowering, fruiting, and leaf-fall (Bennie *et al.*, 2016). These responses are driven by the direct influence of light but also through its indirect actions on physiological processes (Steinlechner, 1996). Natural light is, however, more than just an indicator of time: predictable changes in the direction and angle of the sun and the moon act as a natural compass comparable to the Earth's magnetic fields. The slow movement of the sun, moon, and even stars across the sky, and associated patterns of polarised light in the night sky provide a multitude of celestial means for species to navigate from minute to minute (as they forage or search for mates) or for longer periods of activity such as nightly commuting or seasonal migration (Papi, 1992).

Of course, the efficacy of natural light as clock, calendar, and compass is reduced when it is masked by non-natural, competing light sources. Human deployment of artificial nocturnal lighting—initially through fire then in the various forms of

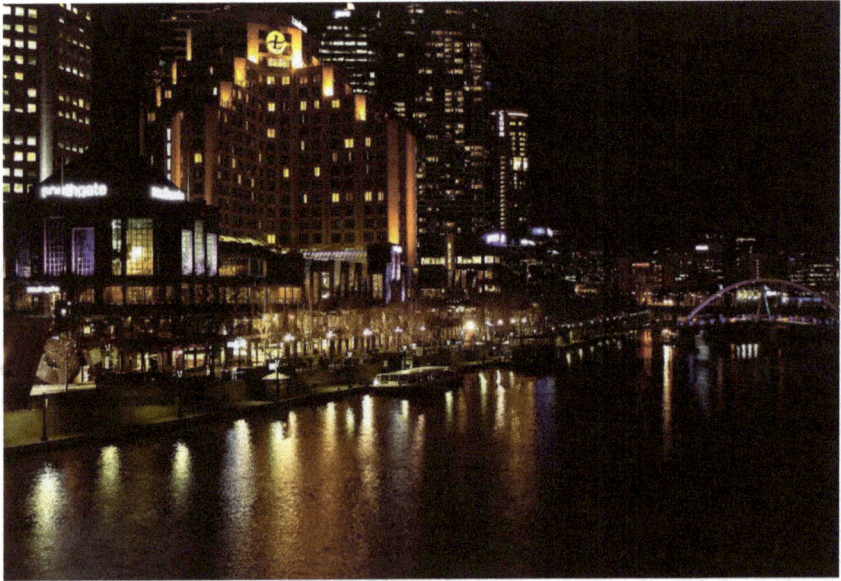

Figure 8.1 A typical city at night—Naarm (Melbourne, Australia)—demonstrating the various forms of artificial night lighting present including the light spills from the point sources of light into the Birrarung (Yarra River). Photograph by Anne Aulsebrook.

Figure 8.2   An example of skyglow. This image was taken from Serendip Sanctuary, a rural location in Victoria (Australia). The visible light is the skyglow from the city of Naarm (Melbourne), nearly 60 km away. Photograph by Joanna Durrant.

outdoor lighting—represents one of the most pervasive anthropogenic stressors on Earth. Globally, approximately 30% of the terrestrial environment (Gaston *et al.*, 2014) and 22% of coastlines (Davies *et al.*, 2020) are affected by some form of ecological light pollution (Figure 8.1). Light pollution comes in many guises. Point sources of light (including streetlights, car headlamps, sports grounds, and other urban structures) directly increase the brightness of the local environment. However, light can also be reflected upward and scattered into the atmosphere, extending tens of kilometres from the original light source in the form of skyglow (Kyba *et al.*, 2012) (Figure 8.2). On cloudy nights, scattered light may even reflect back down to Earth, further increasing the brightness of both local and distant skies.

ALAN masks natural light

As already alluded to, a central problem with ALAN is that it masks and disrupts natural light cycles (Gaston *et al.*, 2014). In artificially lit environments, the transition between day and true night extends beyond natural sunset and commences before natural sunrise; the monthly lunar cycle is dampened because urban lighting can be 10 to 100 times brighter than even the fullest brightest moon; and the almost imperceptible changes in day-length that characterise and define the seasons are less well-defined. Such intrusions to natural light signals inevitably affect the behaviours and physiologies of species living in light-polluted environments.

In practice, the masking presence of ALAN can alter the timing of development and growth, shift mating, foraging, and sleep patterns and disrupt interactions between species.

Artificial nocturnal illumination expands the daytime "photic niche", allowing typically diurnal species, including humans, to commence daytime activities earlier and extend them later into what should be the relative darkness of night. For plants, the presence of artificial light can affect the timing and investment in photosynthesis (Raven and Cockell, 2006); for animals it can increase the amount of time spent searching and consuming food (Santos *et al.*, 2010) or advertising to mates (Dickerson, Hall and Jones, 2022). This can be advantageous, but it may also have costs. For example, light pollution can promote night-time photosynthesis and transpiration in plants (Lockett *et al.*, 2022), but this may lead to fewer flowers and seeds produced (Viera-Perez *et al.*, 2019), earlier spring bud-burst (Ffrench-Constant *et al.*, 2016), delayed autumnal leaf-fall (Matzke, 1936), and shifts in resource investment from roots to leaves (Lockett *et al.*, 2022). Similarly, in animals, brighter nights are beneficial for visually active predators but there are obvious consequences for their prey (Firebaugh and Haynes, 2019; Wilson, 2020). Conversely, expanding daylight hours for diurnal species means an inevitable retraction of the temporal window of the night. This is potentially catastrophic for nocturnal species, particularly light-phobic species, because they have less time to perform their daily activities such as foraging or seeking mates (Bird, Branch and Miller, 2004). Since species respond differently to artificial lighting, disrupting natural light cycles can lead to serious ecological mismatches. For example, for the least horseshoe bat, *Rhinolophus pusillus*, individuals living in urbanised environments delay emergence from their roosts, which means they miss the peak activity period of their insect prey, which can negatively affect their health (Luo *et al.*, 2021).

The observed changes due to the presence of artificial light can be explained in part by the masking or disruption of hormonal (in animals) or other physiological pathways (in organisms as diverse as bacteria, algae, plants, and animals) (Guan *et al.*, 2022; Murphy *et al.*, 2022). For many seasonally breeding birds and mammals, natural increases in day length, associated with the onset of spring, stimulate and synchronise the production of reproductive hormones and the timing of annual reproductive events (Rowan, 1938). In environments that are exposed year-round to ALAN, loss of these subtle shifts in daylength can affect sexual maturation and thus natural patterns of mating (Kempenaers *et al.*, 2010; Da Silva, Valcu and Kempenaers, 2016). A classic example of ALAN affecting hormonal change be seen in the strictly seasonal tammar wallaby (*Notamacropus eugenii*). Under natural light cycles, Tammar wallaby births are tightly synchronised to day-length changes. However, in light-polluted areas, natural light cues are masked, hormones are disrupted, and births are delayed and desynchronised, which may result in a seasonal mismatch between new-born offspring and the availability of the resources they require for their healthy development (Robert *et al.*, 2015).

A final effect, while perhaps not as abundant in nature, but nonetheless critical for the species concerned, is that artificial lighting masks the signals of bioluminescent nocturnal species (Owens and Lewis, 2018). For example, male fireflies, *Photinus obscurellus*, produce light signals to attract females, who respond in kind. In areas illuminated at night, not only do males produce fewer signals, females rarely, if ever, respond. In such species, ALAN disrupts reproduction on a nightly basis, reducing population sizes and potentially placing these species at risk of local extinction (Holker *et al.*, 2010).

The power of ALAN to attract, distract, and repel

A stroll in any urban environment reveals insects flying around streetlamps, birds and bats circling around tall, brightly lit buildings, and potentially a mammal standing motionless and disoriented in the headlamps of a car. Observations of flying animals aggregating around outdoor lighting, both on land and at sea, are littered throughout history. However, why animals respond in this way is not well understood. One possibility is that birds and insects mistake point sources of light such as street lighting for celestial signals like the moon or stars, which they would normally use for navigation (Rich and Longcore, 2006). The consequences of this mistake can be catastrophic. Animals can also become functionally blind and disorientated when exposed to even the dimmest of artificial lighting, which may explain the apparent hypnotising effect of car and streetlights for many mammals (Beier, 2006). Aerial species, such as bats, birds, and insects, may become trapped by a light, flying around until exhausted, until they perish, or the lights are turned off (Rich and Longcore, 2006). At a global scale, bright city lights appear to pull birds off their usual migratory pathways, resulting in them landing in suboptimal environments that do not contain the resources required for reproduction and survival (Cabrera-Cruz, Smolinsky and Buler, 2018). However, scattered and polarised light can also create conflict for animals if it reflects off concrete and other impervious urban surfaces. Such reflected light resembles water, attracting and disorienting insects, such as dragonflies, mayflies, and moths, which ordinarily lay their eggs in water to ensure they do not desiccate (Frank, 2004; Horvath, Kriska and Robertson, 2014). Eggs laid onto urban surfaces are unlikely to survive, restricting the population size of the next generation.

A potential positive effect of lighting for predatory animals is that it promotes the clustering of their prey (such as insects, small mammals, and birds). For predators, lights are thus indicative of a large and accessible meal. The clearest and most abundant examples of predators benefiting from "prey clustering" can be seen in insectivorous bats (Rydell, 2006) and the classic sit-and-wait predators—the spiders (Heiling, 1999; Willmott *et al.*, 2019). For these groups, artificial lights can massively increase foraging rates, with corresponding health and fitness benefits. However, many strictly nocturnal species, including some microbats and other small mammals, are shy or completely photophobic of artificial lighting (Rydell, 2006; Spoelstra *et al.*, 2017). These species actively avoid areas with lights or switch their activity patterns when nearing artificial lights, reducing their ability to feed, and potentially lowering their own health and fitness. Of concern ecologically is that the changes we see at the level of individual species, whether predators or prey, can have cascading effects at the level of the community and even ecosystem (Davies, Bennie and Gaston, 2012; Lockett *et al.*, 2021; Marangoni *et al.*, 2022) and such impacts are evident in both terrestrial and aquatic environments (Manfrin *et al.*, 2017; Mayer-Pinto *et al.*, 2021). Of even greater concern is that some of the light-related changes we observe in nocturnal communities seem to spill over into the day (Davies, Bennie and Gaston, 2012; Knop *et al.*, 2017) and increasing evidence links light pollution with global declines in vertebrate and invertebrate biodiversity (Holker *et al.*, 2010).

ALAN impacts health and fitness

In addition to their visible effects of on activity and behaviour, natural light cycles are integral to the function and control of internal biological clocks and associated chemicals. However, the physical properties of natural and artificial light are comparable, and thus, artificial light can significantly disrupt natural physiological processes. Melatonin, a photo-sensitive biological molecule, provides a stark example. Believed to have arisen during the earliest evolution of life on the planet, melatonin is present throughout most, possibly all, of life (Jones *et al.*, 2015; Grubisic *et al.*, 2019). Typically, melatonin concentrations peak during the night and, due to its sensitivity to light, are at their lowest during the day. While best known for its role as a natural timekeeper (it is integral for stabilising our internal clocks every 24 hours), its second function, and likely evolutionary origin, is as a powerful antioxidant (Zhao *et al.*, 2019). Yet even dim ALAN, less than one tenth the intensity of a typical streetlight, can inhibit the synthesis of melatonin. The presence of light at night is linked with shifts in growth, development and survival in non-human animals, but it is also linked to a suite of health consequences that affect all animals, including humans, such as increased risk of disease, loss of immune function, reduced fertility, disrupted sleep, weight gain, and (in humans) impaired mental health (for a comprehensive review in animals, including humans, see Svechkina, Portnov and Trop, 2020). While night lighting is typically installed for the benefit of humans at the expense of wildlife, these shared health impacts highlight our common biological interest in reducing the scope of artificial light.

One area that is of broad interest in terms of health is the effect of light at night on sleep. The fitness implications arising from loss of sleep are substantial for humans, including reduced cognitive and immune function, mood disorders, and increased weight gain (Cain *et al.*, 2020). Accumulating evidence suggests that this may also hold true for animals (Dominoni, Borniger and Nelson, 2016; Aulsebrook *et al.*, 2018). Experimental evidence on diurnal birds indicates earlier onset of daily activity and increased nocturnal activity in the presence of artificial light (Raap, Pinxten and Eens, 2016; Ouyang *et al.*, 2017; Aulsebrook, Lesku *et al.*, 2020). These findings have been corroborated in laboratory investigations where light at night affects both the quality and quantity of birds' sleep (Aulsebrook, Connelly *et al.*, 2020; Ren *et al.*, 2022). Physiological effects aside, going to sleep when the visual environment is so bright may not be an option for many species because increased brightness means an increased risk of predation (Aulsebrook *et al.*, 2018). Perhaps more worrying is the fact that changes in nocturnal sleeping patterns may tip into the day. Animals can catch up sleep during the day, but this reduces the time they can spend on other activities, such as foraging, mating, or watching for predators (Aulsebrook, Connelly *et al.*, 2020; Aulsebrook, Lesku *et al.*, 2020).

Conflict with human need

Fundamentally, nocturnal illumination fulfils a basic human need: it enhances the visual environment for our diurnally evolved eyes during a time of natural darkness. Activities that were traditionally restricted to daylight hours—work, travel, outdoor recreation, exercise, and socialising—can now occur at any time of the day or night. As a consequence, humans are no longer forced to choose how to spend their hours of precious daylight—work versus sporting activities versus caregiving versus education—and it is now socially and economically viable to extend the time we have available and fit all of it in. Indeed, just about the only activities that are off-limits at night are those that only occur in predominantly unlit areas, such as surfing or bushwalking. However, even some of these can be facilitated with a portable light source (such as a head-torch).

The use of night-time lighting is also not limited to the merely practical (with streetlights and car headlights, for example). The presence of people taking advantage of the illuminated night generates fresh demands for lighting: brightly lit advertising billboards, shop signs and windows, illuminated flags and banners, and other outdoor seasonal lighting. Much like the firefly competing to be seen, the presence of competing light sources drives an upward spiral of lighting intensity: shops and advertisers seek to out-shine the competition and grab attention, while drivers demand ever brighter headlights to see past the blinding lights of oncoming traffic. Ultimately, modern societies are now entirely reliant on ALAN, both culturally and economically, and thus, human needs and desires are entirely (and potentially catastrophically) conflicted with ecological outcomes.

Finding balance

While light pollution is the only form of pollution than can be ended with the flick of a switch, mitigating its ecological effects poses enormous challenges. The moon and stars provide the only night-time light that is guaranteed not to cause ecological harm; thus, the human need or desire for constant bright nights will always conflict with the ecological need for natural darkness. However, while giving up outdoor lighting altogether may be impractical, there is still much we can do to reduce this conflict and its impact on wildlife and natural systems. Ironically, even the most contentious strategies to reduce the impact of artificial light for non-human species could benefit the human animal, because we all suffer some comparable adverse effects from exposure to night-time light. This shared interest provides an important tool for advancing mitigation strategies.

Convincing residents and private landowners to reduce their outdoor lighting use is often a question of education. Most people remain unaware of the ecological harm emanating from their porch lights, indoor lights shining through uncovered windows, or "ambient" garden lighting. However, like all social norms, attitudes to outdoor lighting are amenable to change. In our experience, people from all walks of life are willing to decrease their use of outdoor lighting once made aware of its adverse effects. Wider education regarding the impacts of artificial light on wildlife—and simple steps to mitigate it—should produce an immediate dividend in the form of behaviour change by early adopters which may shift attitudes more broadly.

Reducing publicly owned outdoor lighting requires education of decision-makers and the community, as lighting policies largely reflect public demand for safety, security, ambience, or extended hours of activity. Before seeking to implement change, it is vital to understand where the demand for existing public lighting originated, while recognising that strategies for limiting outdoor lighting may differ widely with context. For example, pedestrian areas and cycle paths used only intermittently at night are often good candidates for adaptive lighting, using sensors and dimmers to provide light only when it is required. Benefits may also be gained from changes to the intensity and colour of light—human eyes require minimal light for wayfinding, about the same as required to read, and bright white lighting increases glare, reduces visibility and ironically makes users feel less safe. Other scenarios, such as sporting facilities (outdoor tennis courts and football grounds), require different solutions, since lighting must be constant and bright enough to see small, fast-moving objects. Reducing ecological impacts in this context requires high-quality focused beams, good shielding, strict limits on hours of activity, and careful preferencing of lighting investment toward facilities away from ecologically important habitat. Public lighting scenarios, including roads, carparks, jetties, marinas, and historic buildings, likewise require context-dependent solutions that must consider the rationale for existing or planned lighting. In short, there is no one-size-fits-all solution.

Public lighting also suffers from agency problems and vested interests that take work to untangle. Ownership and practical control of public lighting assets vary widely within and between jurisdictions and may lie with various levels of government, electricity networks, residential associations, private enterprise, or some combination of these. Understanding who owns and who pays for lighting is an important first step in targeting dark skies advocacy. Agencies footing the bill for lighting are likely to be more receptive to strategies that reduce overall costs than those that profit from existing arrangements. Strategies requiring capital investment, such as adding adaptive lighting technology or changing luminaires, will be more palatable if they simultaneously reduce running costs (for example by dimming lights or reducing their hours of operation). Such strategies have saved local councils in the United Kingdom millions of pounds over the last decade with neutral or beneficial impacts on road safety and crime rates (Green *et al.*, 2015; Steinbach *et al.*, 2015).

The ecological benefits of reduced lighting also need to be weighed against other social and environmental harms. For example, greater participation in organised sport requires sporting facilities to increase their hours of operation, and thus lighting, but also benefits human health and social connectivity. Reducing the hours of lighting in this context may have human costs or encourage land clearing for additional facilities to meet demand—as always, such countervailing considerations are hyper-local and context-specific.

Over and above, public lighting occurs within a legal, regulatory, and social framework that can place limits on the short-term steps that willing actors can undertake. Altering industry standards, codes of practice, laws, and government lighting policies can be slow and frustrating; however, such changes provide the best means of ensuring that future public lighting does not reflect current mistakes. Recent years have seen a flurry of regulations and standards aimed at curbing light pollution—most EU nations, for example, now have binding limits on light pollution at the national or sub-national level (Widmer *et al.*, 2022). Moreover, community support for outdoor lighting, particularly in the road safety and personal security context, is often highly qualified. Research suggests that most people hold nuanced views of lighting, are cognizant of the benefits of dark nights, and are unconcerned by reductions in outdoor light (Green *et al.*, 2015). Even those advocating for increased lighting often do not expect it to reduce fear of crime or increase accessibility and night-time participation (Koskela and Pain, 2000). In summary, progress in making artificial lighting more sensitive to the needs of non-humans and humans requires integration of multiple sectors (personal, public, commercial) and disciplines (ecologists, engineers, lighting designers, administrators) and a recognition that while compromise is necessary, it ultimately benefits both humans and ecosystems.

Mitigating ALAN impacts on non-humans

Often, there is a genuine desire to improve ecological outcomes, but what mitigation might look like is not always clear. Moreover, the effects of different intensities, colours, and timings of artificial light varies widely across organisms, meaning it is unlikely there will be a simple, single solution. Indeed, a lighting strategy that might benefit one species may be costly for another. The placement of lighting in the landscape also determines the extent of ecological impact. For this reason, lighting strategies may need to respond to specific species and/or communities but how this is achieved may vary with context. Readers interested in lighting strategies that aim to minimise impacts on wildlife will benefit greatly from the Australian government's *National Light Pollution Guidelines*, which have now been adopted by the United Nations Convention on Migratory Species (Commonwealth Government of Australia, 2020). At the habitat or landscape scale, the *National Light Pollution Guidelines* recommend the following strategies for reducing the impacts of artificial light on ecological communities:

- Maintaining naturally dark habitat patches and connecting corridors whenever possible.
- Avoiding the creation of "light barriers" that can fragment an intact habitat patch and prevent movement of species within the patch or that can reduce connectivity between neighbouring patches.

- Providing natural or near-natural dark corridors wherever possible.
- Avoiding, removing, redirecting, or shielding artificial lights within and close to habitat patches wherever possible, and keeping intensity as low as practicable, noting that artificial low-intensity light (well below full-moon-light levels) can disrupt terrestrial and aquatic flora and fauna.
- Minimising effects of intermittent mobile light sources, such as vehicle headlights and vessel deck lights.

If lighting must be installed in an ecologically sensitive area, other mitigation measures to consider include the following:

- Using narrow-spectrum, long-wavelength amber or red lighting (although we note that, while this may benefit invertebrates, its effects on other animal groups [fish, birds, amphibians, mammals] is highly variable, and it can disrupt plant physiology).
- Implementing part-night lighting schemes to reduce the duration of artificial light.
- Using motion sensor lighting or dimmers, which may reduce the overall amount of light emitted.

Even where lighting is non-negotiable, a multitude of solutions can shield non-humans from the worst of its effects. These include the use of trees, fencing or landscaping to shade habitat, and planting corridors or islands of dense foliage to provide shelter and assist the movement of small animals across the landscape. Ultimately, due to its entanglement with human desires and decision-making power, the use of ALAN will likely (at best) be minimised. However, if we could aim to achieve even this, we will dramatically benefit all life on Earth, human and non-human alike.

References

Aulsebrook, A., Connelly, F., Johnsson, R., Jones, T., Mulder, R., Hall, M., Vyssotski, A. and Lesku, J. (2020a) 'White and amber light at night disrupt sleep physiology in birds', *Current Biology*, 30, p. 0220035. https://doi.org/10.1016/j.cub.2020.06.085.

Aulsebrook, A., Jones, T., Mulder, R. and Lesku, J. (2018) 'Impacts of artificial light at night on sleep: A review and prospectus', *Journal of Experimental Zoology Part A—Ecological and Integrative Physiology*, 329, pp. 409–418. https://doi.org/10.1002/jez.2189.

Aulsebrook, A., Lesku, J., Mulder, R., Goymann, W., Vyssotski, A. and Jones, T. (2020b) 'Streetlights disrupt night-time sleep in urban black swans', *Frontiers in Ecology and Evolution*, 8(13). https://doi.org/10.3389/fevo.2020.00131.

Beier, P. (2006) 'Effects of artificial night lighting on terrestrial mammals', in Rich, C. and Longcore, T. (eds.) *Ecological consequences of artificial night lighting*. Washington, DC: Island Press, pp. 19–42.

Bennie, J., Davies, T., Cruse, D. and Gaston, K. (2016) 'Ecological effects of artificial light at night on wild plants', *Journal of Ecology*, 104, pp. 611–620. https://doi.org/10.1111/1365-2745.12551.

Bird, B., Branch, L. and Miller, D. (2004) 'Effects of coastal lighting on foraging behavior of beach mice', *Conservation Biology*, 18, pp. 1435–1439.

Cabrera-Cruz, S., Smolinsky, J. and Buler, J. (2018) 'Light pollution is greatest within migration passage areas for nocturnally-migrating birds around the world', *Scientific Reports*, 8(8). https://doi.org/10.1038/s41598-018-21577-6.

Cain, S., McGlashan, E., Vidafar, P., Mustafovska, J., Curran, S., Wang, X., Mohamed, A., Kalavally, V. and Phillips, A. (2020) 'Evening home lighting adversely impacts the circadian system and sleep', *Scientific Reports*, 10, p. 19110. https://doi.org/10.1038/s41598-020-75622-4.

Commonwealth Government of Australia. (2020) *National light pollution guidelines for wildlife including marine turtles, seabirds and migratory shorebirds*. Canberra: Commonwealth of Australia.

Da Silva, A., Valcu, M. and Kempenaers, B. (2016) 'Behavioural plasticity in the onset of dawn song under intermittent experimental night lighting', *Animal Behaviour*, 117, pp. 155–165. https://doi.org/10.1016/j.anbehav.2016.05.001.

Davies, T., Bennie, J. and Gaston, K. (2012) 'Street lighting changes the composition of invertebrate communities', *Biology Letters*, 8(7), pp. 64–767. https://doi.org/10.1098/rsbl.2012.0216.

Davies, T., McKee, D., Fishwick, J., Tidau, S. and Smyth, T. (2020) 'Biologically important artificial light at night on the seafloor', *Scientific Reports*, 10, p. 12545. https://doi.org/10.1038/s41598-020-69461-6.

de Winter, N., Goderis, S., Van Malderen, S., Sinnesael, M., Vansteenberge, S., Snoeck, C., Belza, J., Vanhaecke, F. and Claeys, P. (2020) 'Subdaily-scale chemical variability in a Torreites Sanchezi rudist shell: Implications for rudist paleobiology and the Cretaceous day-night cycle', *Paleoceanography and Paleoclimatology*, 35, p. e2019PA003723. https://doi.org/10.1029/2019PA003723.

Dickerson, A., Hall, M. and Jones, T. (2020) 'The effect of variation in moonlight on nocturnal song of a diurnal bird species', *Behavioral Ecology and Sociobiology*, 74. https://doi.org/10.1007/s00265-020-02888-z.

Dickerson, A., Hall, M. and Jones, T. (2022) 'The effect of natural and artificial light at night on nocturnal song in the diurnal willie wagtail', *Science of the Total Environment*, 808. https://doi.org/10.1016/j.scitotenv.2021.151986.

Dominoni, D., Borniger, J. and Nelson, R. (2016) 'Light at night, clocks and health: From humans to wild organisms', *Biology Letters*, 12, p. 4. https://doi.org/10.1098/rsbl.2016.0015.

Ffrench-Constant, R., Somers-Yeates, Bennie, J., Economou, T., Hodgson, D., Spalding, A. and McGregor, P. (2016) 'Light pollution is associated with earlier tree budburst across the United Kingdom', *Proceedings. Biological Sciences*, 283, p. 20160813. https://doi.org/10.1098/rspb.2016.0813.

Firebaugh, A. and Haynes, K. (2019) 'Light pollution may create demographic traps for nocturnal insects', *Basic and Applied Ecology*, 34, pp. 118–125. https://doi.org/10.1016/j.baae.2018.07.005.

Frank, K. (2004) 'Effect of artificial night lighting on moths', in Rich, C. and Longcore, T. (eds.) *Ecological consequences of artificial light pollution*. Washington, DC: Island Press, pp. 305–344.

Gaston, K., Duffy, J., Gaston, S., Bennie, J. and Davies, T. (2014) 'Human alteration of natural light cycles: Causes and ecological consequences', *Oecologia*, 176, pp. 917–931. https://doi.org/10.1007/s00442-014-3088-2.

Green, J., Perkins, C., Steinbach, R. and Edwards, P. (2015) 'Reduced street lighting at night and health: A rapid appraisal of public views in England and Wales', *Health and Place*, 34, pp. 171–180. https://doi.org/10.1016/j.healthplace.2015.05.011.

Grubisic, M., Haim, A., Bhusal, P., Dominoni, D., Gabriel, K., Jechow, A., Kupprat, F., Lerner, A., Marchant, P., Riley, W., Stebelova, K., van Grunsven, R., Zeman, M., Zubidat, A. and Holker, F. (2019) 'Light pollution, circadian photoreception, and melatonin in vertebrates', *Sustainability*, 11, pp. 136–141. https://doi.org/10.3390/su11226400.

Guan, Q. Y., Wang, Z. X., Cao, J., Dong, Y. L. and Chen, Y. X. (2022) 'The role of light pollution in mammalian metabolic homeostasis and its potential interventions: A critical review', *Environmental Pollution*, 312. https://doi.org/10.1016/j.envpol.2022.12004.

Heiling, A. (1999) 'Why do nocturnal orb-web spiders (Araneidae) search for light?' *Behavioral Ecology and Sociobiology*, 46, pp. 43–49.

Holker, F., Wolter, C., Perkin, E. and Tockner, K. (2010) 'Light pollution as a biodiversity threat', *Trends in Ecology and Evolution*, 25, pp. 681–682. https://doi.org/10.1016/j.tree.2010.09.007.

Horvath, G., Kriska, G. and Robertson, B. (2014) *Anthropogenic polarization and polarized light pollution inducing polarized ecological traps*. Berlin: Springer-Verlag Berlin.

Jones, T., Durrant, J., Michaelides, E. and Green, M. (2015) 'Melatonin: A possible link between the presence of artificial light at night and reductions in biological fitness', *Philosophical Transactions of the Royal Society B-Biological Sciences*, 370, p. 10. https://doi.org/10.1098/rstb.2014.0122.

Kempenaers, B., Borgström, P., Loës, P., Schlicht, E. and Valcu, M. (2010) 'Artificial night lighting affects dawn song, extra-pair siring success, and lay date in songbirds', *Current Biology*, 20, pp. 1735–1739. https://doi.org/10.1016/j.cub.2010.08.028.

Knop, E., Zoller, L., Ryser, R., Erpe, C., Horler, M. and Fontaine, C. (2017) 'Artificial light at night as a new threat to pollination', *Nature*, 548, pp. 206–209. https://doi.org/10.1038/nature23288.

Koskela, H. and Pain, R. (2000) 'Revisiting fear and place: Women's fear of attack and the built environment', *Geoforum*, 31, pp. 269–280.

Kyba, C., Ruhtz, T., Fischer, J. and Holker, F. (2012) 'Red is the new black: How the colour of urban skyglow varies with cloud cover', *Monthly Notices of the Royal Astronomical Society*, 425, pp. 701–708. https://doi.org/10.1111/j.1365-2966.2012.21559.x.

Lockett, M., Jones, T., Elgar, M., Gaston, K., Visser, M. and Hopkins, G. (2021) 'Urban street lighting differentially affects community attributes of airborne and ground-dwelling invertebrate assemblages', *Journal of Applied Ecology*, 58, pp. 2329–2339. https://doi.org/10.1111/1365-2664.13969.

Lockett, M., Rasmussen, R., Arndt, S., Hopkins, G. and Jones, T. (2022) 'Artificial light at night promotes bottom-up changes in a woodland food chain', *Environmental Pollution*, 310, p. 119803. https://doi.org/10.1016/j.envpol.2022.119803.

Luo, B., Xu, R., Li, Y., Zhou, W., Wang, W., Gao, H., Wang, Z., Deng, Y., Liu, Y. and Feng, J. (2021) 'Artificial light reduces foraging opportunities in wild least horseshoe bats', *Environmental Pollution*, 288, p. 117765. https://doi.org/10.1016/j.envpol.2021.117765.

Manfrin, A., Singer, G., Larsen, S., Weiss, N., van Grunsven, R., Weiss, N., Wohlfahrt, S., Monaghan, M. and Holker, F. (2017) 'Artificial light at night affects organism flux across ecosystem boundaries and drives community structure in the recipient ecosystem', *Frontiers in Environmental Science*, 5, p. 14. https://doi.org/10.3389/fenvs.2017.00061.

Marangoni, L., Davies, T., Smyth, T., Rodriguez, A., Hamann, M., Duarte, C., Pendoley, K., Berge, J., Maggi, E. and Levy, O. (2022) 'Impacts of artificial light at night in marine ecosystems—a review', *Global Change Biology*, 28, pp. 5346–5367. https://doi.org/10.1111/gcb.16264.

Matzke, E. (1936) 'The effect of street lights in delaying leaf-fall in certain trees', *American Journal of Botany*, 23, pp. 446–452.

Mayer-Pinto, M., Jones, T., Swearer, S., Robert, K., Bolton, D., Aulsebrook, A., Dafforn, K., Dickerson, A., Dimovski, A., Hubbard, N., McLay, L., Pendoley, K., Poore, A., Thums, M., Willmott, N., Yokochi, K. and Fobert, E. (2021) 'Light pollution: A landscape-scale issue requiring cross-realm consideration', *UCL Open: Environment*. https//doi.org/10.14324/111.444/000103.v1.

Mougeot, F. and Bretagnolle, V. (2000) 'Predation risk and moonlight avoidance in nocturnal seabirds', *Journal of Avian Biology*, 31, pp. 376–386.

Murphy, S., Vyas, D., Sher, A. and Grenis, K. (2022) 'Light pollution affects invasive and native plant traits important to plant competition and herbivorous insects', *Biological Invasions*, 24, pp. 599–602. https://doi.org/10.1007/s10530-021-02670-w.

Ouyang, J. Q., de Jong, M., van Grunsven, R., Matson, K., Haussmann, M., Meerlo, P., Visser, M. and Spoelstra, K. (2017) 'Restless roosts: Light pollution affects behavior, sleep, and physiology in a free-living songbird', *Global Change Biology*, 23, pp. 4987–4994. https://doi.org/10.1111/gcb.13756.

Owens, A. C. S. and Lewis, S. M. (2018) 'The impact of artificial light at night on nocturnal insects: A review and synthesis', *Ecology and Evolution*, 8(22), pp. 11337–11358. https://doi.org/10.1002/ece3.4557

Papi, F. (1992) *Animal homing*. London: Chapman and Hall.

Raap, T., Pinxten, R. and Eens, M. (2016) 'Artificial light at night disrupts sleep in female great tits (*Parus major*) during the nestling period, and is followed by a sleep rebound', *Environmental Pollution*, 215, pp. 125–134. https://doi.org/10.1016/j.envpol.2016.04.100.

Raven, J. and Cockell, C. (2006) 'Influence on photosynthesis of starlight, moonlight, planetlight, and light pollution (reflections on photosynthetically active radiation in the universe)', *Astrobiology*, 6, pp. 668–675.

Ren, Z. F., Chen, Y. Q., Liu, F. B., Ma, X. L., Ma, J. X. and Liu, G. (2022) 'Effects of artificial light with different wavelengths and irradiances on the sleep behaviors of chestnut buntings (*Emberiza rutila*)', *Biological Rhythm Research*, 53, pp. 1454–1473. https://doi.org/10.1080/09291016.2021.1958542.

Rich, C. and Longcore, T. (2006) *Ecological consequences of artificial night lighting*. Washington, DC: Island Press.

Robert, K., Lesku, J., Partecke, J. and Chambers, B. (2015) 'Artificial light at night desynchronizes strictly seasonal reproduction in a wild mammal', *Proceedings of the Royal Society B-Biological Sciences*, 282, p. 20151745. https://doi.org/10.1098/rspb.2015.1745.

Rowan, W. (1938) 'Light and seasonal reproduction in animals', *Biological Reviews of the Cambridge Philosophical Society*, 13, pp. 374–402.

Rydell, J. (2006) 'Bats and their insect prey at streetlights', in Rich, C. and Longcore, T. (eds.) *Ecological consequences of artificial night lighting*. Washington, DC: Island Press, pp. 43–61.

Santos, C., Miranda, A., Granadeiro, J., Lourenco, P., Saraiva, S. and Palmeirim, J. (2010) 'Effects of artificial illumination on the nocturnal foraging of waders', *Acta Oecologica-International Journal of Ecology*, 36, pp. 166–172. https://doi.org/10.1016/j.actao.2009.11.008.

Spoelstra, K., van Grunsven, R., Ramakers, J., Ferguson, K., Raap, T., Donners, M., Veenendaal, E. and Visser, M. (2017) 'Response of bats to light with different spectra: Light-shy and agile bat presence is affected by white and green, but not red light', *Proceedings of the Royal Society B-Biological Sciences*, 284, p. 8. https://doi.org/10.1098/rspb.2017.0075.

Steinbach, R., Perkins, C., Tompson, L., Johnson, S., Armstrong, B., Green, J., Grundy, C., Wilkinson, P. and Edwards, P. (2015) 'The effect of reduced street lighting on road casualties and crime in England and Wales: Controlled interrupted time series analysis', *Journal of Epidemiology and Community Health*, 69, p. 1118. http://dx.doi.org/10.1136/jech-2015-206012.

Steinlechner, S. (1996) 'Melatonin as a chronobiotic: PROS and CONS', *Acta Neurobiologiae Experimentalis*, 56, pp. 363–372.

Svechkina, A., Portnov, B. and Trop, T. (2020) 'The impact of artificial light at night on human and ecosystem health: A systematic literature review', *Landscape Ecology*, 35, pp. 1725–1742. https://doi.org/10.1007/s10980-020-01053-1.

Tierney, S., Friedrich, M., Humphreys, W., Jones, T., Warrant, E. and Wcislo, W. (2017) 'Consequences of evolutionary transitions in changing photic environments', *Austral Entomology*, 56, pp. 23–46. https://doi.org/10.1111/aen.12264.

Viera-Perez, M., Hernandez-Calvento, L., Hesp, P. and Santana-del Pino, A. (2019) 'Effects of artificial light on flowering of foredune vegetation', *Ecology*, 100. https://doi.org/10.1002/ecy.2678.

Widmer, K., Beloconi, A., Marnane, I. and Vounatsou, P. (2022) *Review and assessment of available information on light pollution in Europe*. Kjeller: European Environment Agency.

Willmott, N., Henneken, J., Elgar, M. and Jones, T. (2019) 'Guiding lights: Foraging responses of juvenile nocturnal orb-web spiders to the presence of artificial light at night', *Ethology*, 125, pp. 289–297. https://doi.org/10.1111/eth.12852.

Wilson, N. (2020) 'Predator-prey dynamics altered by light pollution', *Frontiers in Ecology and the Environment*, 18, pp. 538–538. https://doi.org/10.1002/fee.2280.

Zhao, D., Yu, Y., Shen, Y., Liu, Q., Zhao, Z. W., Sharma, R. and Reiter, R. (2019) 'Melatonin synthesis and function: Evolutionary history in animals and plants', *Frontiers in Endocrinology*, 10, p. 16. https://doi.org/10.3389/fendo.2019.00249.

9 Preserving darkness in the wildwood

Kimberly Dill

Introduction

Global development serves as a harbinger of ever-increasing quantities of artificial light at night (ALAN). The ubiquity of high-pressure sodium bulbs and bright, white LEDs cast a skyglow that extends artificial lighting beyond the concrete bounds of urban and suburban spaces. Problematically, this *light pollution* hinders human psycho-physiological wellbeing, contributes to the loss of aesthetic, intellectual, and cultural values, and inhibits the flourishing of biodiverse organisms and environments (Dill, 2021; Stone, 2018).

Herein, I utilise an environmental ethical framework to analyse how the loss of natural darkness negatively hinders the conservation and restoration of biodiverse *forested* ecosystems, while simultaneously damaging the irreplaceable relationships that we *Homo sapiens* form with them. In addition, as ALAN inhibits a forest's ability to sequester carbon effectively, it thereby contributes to overall climate change (including extreme shifts in weather and rising global temperatures). To explicate these claims, I employ the key philosophical concepts of *wildness* (defined as a lack of toxic, anthropogenic influences below a vague threshold), *eudaemonia* (flourishing), and *relational value* (according to which the wellbeing of both people and forests is inherently intertwined) (Dill, 2021). Altogether, I contend that the preservation of natural darkness should be conceived of as fundamental to forest, human and global health.

Relationalism: a brief history

Ethics and the environment

Broadly, I contextualise and analyse claims about the necessity of dark sky conservation from within a philosophical, ethical framework. More specifically, light pollutant harms to biodiverse, including forested, species, I argue, are ethically problematic both *intrinsically* (that is, in virtue of their value-in-themselves) and *relationally* (that is, insofar as their presence or absence contributes either to the flourishing or declination of their larger, biotic community). This stripe of ethical *relationalism* attributes a kind of value to more-than-human beings which

DOI: 10.4324/9781003408444-13
This chapter has been made available under a CC-BY-NC-ND 4.0 license

transcends yet integrates both intrinsic and instrumental reasons that motivate care. To understand this claim, it is worth noting that relationalism is motivated by two theoretical components: one descriptive (that is, metaphysical and ontological) and the other normative (that is, ethical). Descriptively, relationalism involves a commitment to the idea that life is best understood interdependently—as fundamentally intertwined, interwoven, and entangled (both climatically and at the level of ecosystems). Normatively, relationalism posits that we ought to act in ways that reciprocally maximise the wellbeing of those with whom we stand in relationship (through, for example, acts of care).

Relationalism's ontological and ethical commitments

Today, some biologists, mycologists, and forest ecologists (Kimmerer, 2013; Margulis, 1967; Sheldrake, 2020; Simard, 2021) have argued that the serious consideration of cooperative *symbiosis* necessitates the rejection of biological and ontological atomism—the view that individuals and species are distinct and wholly separable. One consequence of this view is that the identity or definition of an individual organism by necessity (though not completely) refers to organisms or environmental features outside of it. As organisms cannot be understood in isolation, our biotic world is fundamentally and inextricably *entangled* at its roots (Dill, 2021; Sheldrake, 2020). Relationalism thereby takes seriously Ernst Haeckel's (1866) claim that ecology requires a serious "study [of] the relationships between organisms and their environments: both the places where they live and the thicket of relationships that sustain them" (Sheldrake, 2020, p. 71).

This ontological and biological interdependence firmly suggests that we, *Homo sapiens*, also find ourselves inextricably embedded within a web of biotic relationships. Furthermore, this fact, that which Norwegian philosopher and father of deep ecology, Arne Naess (1973), termed the "relational, total-field image", suggests constraints on and recommendations for certain kinds of behaviour. In order to motivate this argumentative move, it is worth clarifying that relationships can be understood along two dimensions: both descriptively (ecologically) and normatively. In this latter sense, relationships at times bear aesthetic or ethical value and so, by extension, carry normative or moral force.

Drawing on the Aristotelian concept of *eudaemonia*—flourishing—I argue that relationships (in addition to individuals) exhibit degrees of more or less wellbeing. To clarify, Aristotle defined *eudaemonia* as the *telos* or ultimate goal toward which all living beings ultimately aspire. A eudaemonistic life is, in short, a life-well-lived: a flourishing life and the source of all goodness. Good actions or objects thereby acquire their value only derivatively, insofar as they instantiate, or fail to instantiate, eudaemonia effectively. The form of relationalism that I endorse here extends this Aristotelian conception of eudaemonia from individuals to *relationships* and *communities*, conceived of holistically. Different kinds of relationships bear and exhibit differing kinds or degrees of value, for they can be healthy and reciprocally rewarding or may at times—and quite tragically—fall irreparably ill.

Anishinaabe author and environmental scientist Robin Wall Kimmerer (2013) suggests that loving or healthy relationships by necessity exhibit features of care, respect, and reciprocity. In other words, healthy relationships contribute to the *mutual* flourishing of the *relata* who together comprise them. Understood within this broader ethical framework, healthy relationships of this sort are (what I have elsewhere termed) *synergistic* and regulative (Dill, 2021). Those that detract from the wellbeing of one or more of their *relata*—as a direct result of neglect, abuse or other harms perpetuated by one participant are, by contrast, toxic (Dill, 2021). Altogether, the application of the Aristotelian conception of eudaemonia to relationships and individuals recommends the following prescriptive claim: ethical actions are those that contribute to relational flourishing, while actions that detract from reciprocal wellbeing are, by contrast, morally bereft. We are, thereby, morally obliged to act in ways that demonstrate reciprocal care, in gratitude for the rich abundance that we receive in turn.

Extending relationalism to the more-than-human world

Historically, elements of relationalism have been written into many diverse environmental ethical worldviews, including Norwegian deep ecology (Naess, 1973), Anishinaabe traditional ecological knowledge (Kimmerer, 2013), Daoism (Chinn, 2013), the virtue ethics of Confucianism (Hourdequin and Wong, 2005), Aldo Leopold's land ethic (1949), some forms of eco-feminism (Cuomo, 2002), and some Indian environmental philosophy (Shiva, 1988).

Despite this rather broad taxonomic umbrella, relationalists are nevertheless united by the aforementioned set of descriptive and normative core commitments:

1. *Descriptively*: Given the ecological evidence, *Homo sapiens* are inextricably embedded within—and are not fundamentally distinct from—our more-than-human world.
2. *Ethically*: In virtue of the ecological relationships on which we fundamentally rely, *Homo sapiens* are ethically obliged to act in ways that reciprocally contribute to the flourishing of our more-than-human kin.

In order to justify (2), it is worth emphasising that members of our species *Homo sapiens* existentially rely on the material, psychological, emotional, and spiritual sustenance afforded by more-than-human, ecological beings. At the most basic level we are,

> mere heterotroph[s], feeder[s] on the carbon transmuted by others. In order to live, I must consume. That's the way the world works, the exchange of a life for a life, the endless cycling between my body and the body of the world.
>
> (Kimmerer, 2013)

Our flourishing is thus only made possible through exchanges with the complex web of socio-ecological relationships that together constitute our world. Of course, our knowledge about these relationships, facts about whether we are in them at all, and their corresponding ethical obligations may appear to us as, at times, epistemically opaque, for though all *Homo sapiens* necessarily depend on more-than-human flora and fauna for our sustenance (through the agricultural products we consume, textiles we wear, and materials that structure our homes), not all of us are consciously aware that we do so. Nevertheless, and in light of this fact, Kimmerer suggests an extension of the moral sphere to embrace members of our more-than-human world under its purview. In sum, those actions are good which reciprocally contribute to the wellbeing or flourishing of *both* human and more-than-human beings.

"The Woods Are Lovely, Dark, and Deep" (Frost, 1923)

Human relationships with forests

Within the context of this piece, I am particularly interested in remediating pre-existing relationships between human beings and our sylvan kin. After all, our very lives *qua* human beings are entangled with and made possible through our ancestral and contemporary relationships with the plants, mycelia, soils, and more-than-human animals that compose forested ecosystems.

For the sake of brevity, I will not deeply explicate the multiple ways that we benefit from our ongoing relationships with forests. I will, however, briefly gesture at and make reference to some additional, irreplaceable sylvan gifts: from forests, we enjoy a vast range of building materials for shelter such as wood, as well as foodstuffs and priceless medicinal compounds. Existentially, and most pressingly, old-growth groves provide our Earth with approximately one-third of its overall oxygen (Gilhen-Baker, Roviello and Beresford-Kroeger, 2022), whilst forest floras purify toxins from industrially polluted air (Beresford-Kroeger, 2018).

Psycho-physiologically, the immersive, proprioceptive practice of *shinrin yoku* (forest bathing)—in which subjects inhale phytoncides, the essential oils emitted from trees, such as Japanese cedar (*Cryptomeria*) and Hinoki cypress (*Chamaecyparis obtuse*)—proves demonstrably efficacious. More specifically, these biogenic volatile compounds boost immune-system functioning (Ikei, 2015; Li *et al.*, 2007) and the growth of anti-cancerous natural killer and T-cells (Li *et al.*, 2007). Intriguingly, the available psychological research also indicates that the experiential fascination induced in subjects exposed to biodiverse and very large tree varietals measurably alleviates symptoms associated with anxiety, depression, ADHD, and autism spectrum disorder (Kaplan, 1995).

Beyond these beneficial effects, forests provide subjects with opportunities to engage in mindful, auditory listening (for instance, to frog and bird song and stargazing) while facilitating the preservation and cultivation of cultural mythos that are contingent on natural darkness. Altogether, it is therefore clear that sylvan groves bear an irreplaceable transformative and restorative power, for immersion in them is psycho-physiologically healing along multiple dimensions.

A gift for a gift: restoring natural darkness

Given the wealth of material, psychological, spiritual, emotional, and restorative resources afforded to us by forests, reciprocity suggests that we have a strong relational duty to return their generosity with a gift. The particular recommendation that I here have in mind is by no means flashy or complex. Rather, I suggest that we ought to work toward the restoration or conservation of natural darkness in sylvan ecosystems. Though at times seemingly intangible, forests and their biodiverse constituents *require* natural darkness in order to flourish. Their effective re-darkening therefore serves to "sustain the ones who sustain us, [such that they] will last forever" (Kimmerer, 2013, p. 183).

Dwindling, ecological darkness

Increasingly, white and high-pressure illumination disrupts the flourishing of a range of species, from pollinators and nocturnal mammals to trees and their diverse groves. Consequently, one of the great losses associated with dwindling natural darkness is artificial light's detrimental impacts on the functioning and wellbeing of forested ecosystems. The flourishing of forest dwellers is impeded, more specifically, for artificial light disrupts circadian rhythms, hinders effective navigation, and removes the cover afforded by natural darkness that some animals require to avoid predation. As the vast majority of invertebrates (60%) and mammal species (an incredible 69%) are nocturnal, the preservation or restoration of biodiverse flora and fauna *requires* the preservation (or restoration) of natural darkness (Bennie *et al.*, 2016; Hölker *et al.*, 2010).

 To illuminate this claim in a useful and measured way, I will spend some time evaluating how light pollution negatively affects a variety of forest-dwelling species: frugivorous and insectivorous (fruit- and insect-consuming) bats, invertebrates (including insects), diverse flora (including trees), and fungi. Curiously, while the effects of artificial lighting on bat species have been extensively covered in the empirical literature, the effects of artificial lighting on trees and fungi in non-urban settings have not. Accordingly, my analysis of ALAN's detrimental effects focuses on the interdependent interactions betwixt and between the diverse species that dwell therein. More specifically, I analyse how these species positively—and, at times, essentially—contribute to the entangled flourishing of the forested ecosystems they call home. The loss of any one species thereby constitutes a significant harm to the wildwood. Altogether, as we *Homo sapiens* owe a significant debt in reciprocity for the abundance provided by these sylvan zones, we are ethically obligated to work toward their effective re-darkening.

Under cover of darkness: bats, seed dispersal, and reforestation

This exploration begins with an examination of the harmful effects of night-time artificial lighting on one nocturnal, flying mammal, which has been culturally associated with metaphorical and literal darkness. Though at times unnecessarily feared (through e.g., their portrayal in Bram Stoker's *Dracula*), bats play an important ecological role in the creation and maintenance of tropical forests. Serving as seed-dispersing stewards, pollinators, and insect suppressors (Brasileiro, Machado and Aguiar, 2022), bats are indispensable members of and contributors to the health of the wildwood. Light-pollutant harms to bats thus constitute a series of irreparable harms which threaten the wellbeing of the forests in which they dwell.

Given their nocturnal status, bats are particularly sensitive to light. They prefer a cover of darkness *so thick* that even the natural illumination of our moon reduces their hunting and foraging activities (Morrison, 1978; Fleming, 1988). Indeed (and perhaps counter-intuitively), insectivorous bats rarely maximise the potential afforded by artificial lights, which tend to attract a range of invertebrate insects into their glowing field (Stone, Jones and Harris, 2009, 2012). Instead, bats' extreme light shyness prompts them to relinquish these veritable banquets in favour of maintaining an obscured sense of safety. Beyond these, the changes induced by ALAN are vast, including "a delay to leave the nest, decreased sexual activity, changes in flight speed and paths (trajectory, height) as well as significant increases in collisions (~25%) in the presence of lit obstacles" (Falcón *et al.*, 2020).

Fruit- and nectar-consuming bats are likewise affected by ALAN. Most avoid well-lit areas, for illumination also renders them vulnerable to predation (Lowery, Blackman and Abbat, 2009). This is particularly worrying, for frugivorous bats provide an array of irreplaceable ecosystem services to the forests in which they dwell. More specifically, frugivorous bats (including *Leptonycteris curasoae*) pollinate flowers and disperse seeds, contributing to the growth and flourishing of a diversity of plant varietals (Ghanem and Voigt, 2012). Even more pressingly, frugivorous bats can withstand habitat disturbances produced when tropical forests are culled to make way for agriculture or livestock grazing. These regions are often later abandoned by farmers due to factors such as soil depletion, so these regions are subject to devastating and alarmingly high rates of biodiversity loss. Given their role as the primary seed-dispersers of pioneering (i.e., first-wave) plant varietals (Muscarella and Fleming, 2007), neotropical frugivorous bats spur the healthy regeneration and reforestation of depleted ecological groves.

However, given their light-shyness, frugivorous bats *require* the cover afforded by natural darkness to successfully produce their "copious seed rain . . . in deforested habitats" (Lewanzik and Voigt, 2014). ALAN thus hinders reforestation and spurs further biodiversity loss by dissuading frugivorous bats from engaging in their integral stewardship efforts.

Artificial light's effects on plant and tree species

Perhaps surprisingly, few studies have been conducted on how ALAN affects plants in non-urban or suburban settings. Due to variances in lighting intensities (compare, say, the diffuse illumination of skyglow to the direct illumination shone by white-LED streetlamps), it has been difficult to isolate and measure the particular effects of different lighting kinds, levels, and intensities. Instead, the bulk of research has so far focused on artificial light's effects on either plants in urban settings or more broadly, on plant-pollinator relationships (Giavi, Fontaine and Knop, 2021). With this in mind, Bennie *et al.* (2016, p. 612) call on ecologists to "define ecologically meaningful measures of artificial light in the natural environment [in order to] develop understanding of the thresholds and dose-response relationships of light-sensitive processes in plants".

It is clear, nevertheless, that biological beings in general—from bats and human beings to bees *and* trees—have evolved circadian rhythms that track transitions from day to night. More precisely, circadian rhythms serve as internal clocks, which regulate an organism's sleep-wake cycles, influence their behaviour, and determine overall metabolic growth.

Briggs (2006) provides a thorough and compelling review of the available research into the physiological effects of artificial light on the circadian rhythms of urban and suburban plant varietals. Altogether, the empirical evidence summarised demonstrates that artificial light has measurable and sometimes detrimental effects on the season-relative timing of leafing, bud-burst, and flowering in trees and plants more generally, which is (under naturally dark conditions) determined relative to day length and seasonality (Falcón, 2020). In other words, as artificial lighting mimics well-lit days and emulates longer, brighter seasons, it prompts plants to bud earlier and lose their leaves later in the year. This is worrisome, for shifts in seasonal cycles could result in the potential desynchronisation between plants and the pollinator populations on which they rely for successful reproduction, dispersal, and gene flow.

Last but certainly not least, plants, like humans, need rest. Natural darkness provides them with the proper conditions under which to recover from physiological stress and strain (Futsaether *et al.*, 2009). This is crucial, for rest stimulates successful carbon metabolism through photosynthesis, respiration, and photorespiration, which enables plants to synthesise sunlight into food, produce oxygen, and sequester carbon (Gaston *et al.*, 2017). By contrast, the strain of enduring continuous artificial lighting inhibits this process by inducing necrosis (cell injury) and chlorosis (the yellowing of leaves) (Velez Ramirez *et al.*, 2011).

Artificial light, moths, and night-blooming flowers

The bulk of our planet's biodiversity is comprised of invertebrates, including a mind-numbing variety of insect species. Familiarly, insects are highly attracted to artificial illumination, for it emulates the white glow of luminous moonlight. This is problematic, for in some cases it renders insects vulnerable to predation by owls, reptiles, spiders, and amphibians. Highly reflective surfaces only magnify this concern, for their twinkle attracts pollinators (including moths) and disrupts the lifecycles of aquatic insects by drawing them out of and away from water (Bruce-White and Shardlow, 2011). As this chapter seeks to recommend strategies that contribute to the flourishing of the wildwood, I evaluate the threats posed by ALAN to nocturnal, *pollinating insects* in particular. After all, the flourishing of flowering plants, including their successful reproduction and the maintenance of their genetic diversity, requires the pollinating labours of moths, bees, nectarivorous bats, and flies. Indeed, nectarivorous insects are responsible for the bulk of this entomophily.

One nocturnal species is primarily responsible for the pollination of night-blooming flowers: the ethereal, illustrious, and elusive moth (Lepidoptera). Over the course of their adult lifecycles, the relationship between flora and moth grows and deepens, for the nectar that once served as their primary source of suste-nance then eventually feeds their young (MacGregor *et al.*, 2015). Reciprocally, the flourishing of white varietals with sweet, heavy fragrances, including jasmine (*Jasminum*) and bog orchids (*Platanthera leucostachys*), require the distributive ecosystem services afforded by moths. Indeed, from tropical rainforests to conifer-ous forests, moths, including Sphingidae, Noctuidae, and Geometridae, contribute indispensably to the wellbeing and functioning of the complex, varied, and entan-gled ecosystems in which they dwell (Winfree, Bartomeus and Cariveau, 2011). Accordingly, their loss would constitute an irreparable series of harms to biodi-verse, night-blooming plants.

Tragically, nocturnal moth populations have declined by an overwhelming *two-thirds* since the early 1970s (Fox *et al.*, 2013). In addition to climate change and habitat degradation, the available evidence strongly suggests that ALAN may be driving population decreases in Great Britain, the Netherlands, and Finland (Hölker *et al.*, 2010; Fox, 2013; MacGregor *et al.*, 2015). Like other nocturnal ani-mals, ALAN affects moth populations along a few dimensions. Illumination poses a threat to nocturnal moths, for it both renders them highly vulnerable to predation and impedes their capacity to reproduce successfully. Even "low levels of artificial light [have been shown to] inhibit the release of sex pheromones in female moths of the Geometridae species" (Sower, Shorey and Gaston, 1970), while simultane-ously disrupting the healthy development and growth of moth larvae. Furthermore, moth navigation is partially driven by their sensitive vision, a function of their combined simple and compound eyes (Frank, 2006). As bright light obscures the UV markings on flowers, these nectar-foragers are thereby unable to successfully locate and land on their preferred sources of sustenance (Davies, Bennie and Inger, 2013). By extension, ALAN hinders their successful pollination efforts. In some cases, it delays their nocturnal, foraging activities altogether (for, long after the sunsets, moths continue to sense their still-bright world as "still day").

Given their integral ecological role, ALAN's deleterious effects on moth species are potentially momentous, detracting from the functional wellbeing and function-ing of their biodiverse ecosystems (MacGregor *et al.*, 2015). The loss of nocturnal moths thereby constitutes a significant harm to the flourishing of the wildwood.

ALAN, mycelial health, and water pollution

Fungi play a variety of crucial roles in forested ecosystems. In their capacity as decomposing saprobes, they indispensably recycle litter (including leaves, wood, and needles) into nutrient-dense soil. Fungi, furthermore, participate in a complex web of nutrient and chemical exchanges with diverse tree species. For example, mycorrhizal fungi carry excess nitrogen from alder to pine trees through an interwoven and mutualistic underground network. Alder, more specifically, are especially nitrogen-rich, yet in a spectacularly cooperative fashion, contribute this excess abundance to their nitrogen-poor plant neighbours, including pine trees (Simard, 2021). The exchange between alder and pine, however, requires an intermediary, a role fulfilled and facilitated by the mycorrhizal fungi that colonise the roots of both species. In terms of their reciprocal flourishing, fungi benefit from their participation in this mutualistic, forested exchange through the carbohydrates that they in turn receive. Thus, this symbiotic process contributes to the growth and *eudaemonia* of all three, entangled beings.

Relatively little empirical work has been conducted on the relationship between increases in ALAN and forest-dwelling fungi. Nevertheless, we can (with a reasonably high degree of accuracy) infer its effects in the proverbial wild, for ALAN demonstrably stunts the growth of fungal fruiting bodies such as mushrooms and mycelia, while simultaneously impeding fungi's capacity to engage in efficacious litter decomposition (Pu *et al.*, 2020). Controlled laboratory studies have shown, for example, that light hinders mycelial growth in jack-o'-lantern (*Omphalotus olearius*), bitter oyster (*Panels stipticus*), honey fungus (*Armillaria mellea*), and *Mycena citricolor* mushrooms. More precisely, Weitz *et al.* (2001) observed that all four species underwent a decreased period of growth when placed under (artificially bright) light for a 24-hour period. Worryingly, *A. melea* failed to produce rhizomorphs (that is, hyphae strands) under *any* artificial light whatsoever. By contrast, a 24-hour period of total darkness proved optimal for mycelial growth across all four species. Generalising from these studies, it is reasonable to infer that ALAN will have measurable (and potentially detrimental) effects on the growth of wild fungi in forested settings, too.

In addition to stunting fungal growth, the presence of ALAN affects and alters "fungal community composition and the correlations between fungi species" while simultaneously over-stimulating or inhibiting their capacity to engage in efficacious litter decomposition (Pu *et al.*, 2020). Furthermore, when fungi are overly stimulated by artificial light, they engage in the hyper-decomposition of litter. As a result, toxic quantities of arsenic (Pu *et al.*, 2020) and cadmium are released into aquatic streams. In one study, ALAN stimulated fungal litter decomposition to such an extent that cadmium toxicity in nearby aquatic stream-beds increased by an overwhelming 71% (Liu *et al.*, 2020). The effects of this process are, of course, deleterious, for "fungi . . . play key roles in the carbon and nutrient dynamics of stream ecosystems" yet "are more sensitive to pollutants than bacteria" (Liu *et al.*, 2020).

Altogether, the presence of artificial lighting clearly affects fungal, forested communities along (at least) two dimensions: it hinders their healthful growth and impedes their capacity to engage in the efficacious and measured decomposition of forest detritus. Given the crucial role that fungi play in forested ecosystems, harms to fungi constitute a series of both direct and indirect harms to forests, conceived of as wholes.

Conclusions

Ecological networks and entangled flourishing

Throughout this conceptual foray into the wildwood, assessments of diverse beings have made explicit reference to the indispensable and entangled ecosystem services that each species provides. These community structures, including "the occurrence (and frequency) of interactions between species, such as plants and pollinators", are best described as *ecological networks* (Montoya, Pimm and Solé, 2006). In addition, their robustness or resiliency refers to each network's ability to withstand drastic changes or perturbations to their overall composition, through, for instance, extirpation or species loss (Evans, Pocock and Memmott, 2013).

Drivers of ecological change, including global warming and dramatic increases in light pollution, have measurable effects on network resiliency. When, for example, considering the important networks that emerge from interrelationships between plants and pollinators, it is clear that the loss of one has dramatic and, at times, irreparable consequences on the flourishing of the other.

As an isolated phenomenon, the effects of light pollution on ecological networks has not yet been holistically and explicitly studied (MacGregor *et al.*, 2015). Nevertheless, the available empirical data *does* clearly demonstrate that nocturnal moths, plant and tree populations, mycelia, and insectivorous and frugivorous bats are threatened by expanding urbanisation and the artificial light-at-night that it heralds. It is thus reasonable to infer that ALAN's detrimental impacts on the species outlined herein will, by extension, spur a set of cascading consequences for the ecological networks in which they dwell. In sum and most pressingly, the flourishing of the wildwood is threatened by ever-increasing quantities of ALAN.

Darkness and the mitigation of climate change

Interwoven into this analysis has been an implicit gesture toward the climate impacts of light pollution, an indirect yet problematic product of its detrimental effects on biodiverse, forested ecosystems. More specifically, forests demonstrably serve as carbon sinks, for they lock overabundant atmospheric CO_2 into the trees, grasses, shrubs, and mycelial networks that together comprise them. Old-growth forests account for about 10% of carbon sequestration globally (Gilhen-Baker, Roviello and Beresford-Kroeger, 2022) and rely on their healthy mycelial networks in order to flourish. However, the available empirical evidence strongly suggests that ALAN hinders both the mycelial and radial (fruiting-body) growth of fungi. As such, ALAN also poses a threat to the continued growth of old-growth trees while simultaneously preventing fungi from fulfilling their carbon-sequestering role.

We have seen, in addition, that an overabundance of artificial lighting in the tropics dissuades nocturnal keystone species, including bats, from engaging in their efficacious pollination and seed-dispersal efforts. As a result, deforested regions, more specifically those that have been clear-cut to make way for mono-agriculture plantations, are slower to regenerate through fresh growth. This implies that light pollution also *indirectly* contributes to climate change, for it hinders the regeneration of forests which under naturally dark conditions would be successfully seeded and could thereby serve as efficacious carbon sinks.

A relational duty to re-darken forests

Altogether, relationalism recommends a set of duties: epistemically to learn about and ethically to invest in the renewal of healthy, regulative, and synergistic ecological relationships. The particularities of environmental ethical prescriptions should take the contextual features of the flora and fauna that are endemic to each sylvan space into account, for their varying needs will motivate varying constraints on anthropogenic behaviour. Nevertheless, one recommendation can be generalised across forested ecosystems: in reciprocal gratitude for the multiplicity of sylvan gifts received, we *Homo sapiens*, quite simply, are ethically obligated to restore their natural darkness.

Practically, this can be achieved by diverse measures including employing only soft, warm-toned, motion-activated, and low-pressure sodium bulbs, constructing widespread and well-networked wildlife corridors, and devising local policies that constrain lighting trespass near sylvan zones (Dill, 2021; Lewanzik and Voigt, 2014). The construction of interconnected and naturally dark wildlife corridors is particularly efficacious, for they exhibit low degrees of toxic anthropogenic influence and consist of "interspersed [and] peri-urban spaces, which blend [the] barriers between developed and more-than-human realms" (Dill, 2021, p. 16). Even when relatively small, these regions demonstrably allow for the unimpeded migration of more-than-human beings, including bats, between fragmented afforested zones. Furthermore, as darkness corridors utilise motion-activated lighting along their borders, their construction ensures that illuminative trespass remains minimal and limited. Thus, by preserving or re-instating natural darkness in these cordoned-off regions, we will by extension successfully *re-wild* them.

All things told, the cultivation of relational *eudaemonia* between humans and forested ecosystems requires the implementation of wise and considerate boundaries, which ultimately contribute to the wellbeing of those complex eco-relationships in which we find ourselves embedded. Toward this end, we are called on to mindfully sacrifice our diurnal preconceptions of comfort and a global addiction to hyper-illumination in favour of the deep, dark, and flourishing *wild*wood.

References

Bennie, J., Davies, T., Cruse, D. and Gaston, K. (2016) 'Ecological effects of artificial light at night on wild plants', *Journal of Ecology*, 104(3), pp. 611–620.

Beresford-Kroeger, D. (2018) 'The medicine of trees', The Ninth Haig Brown Memorial Lecture, Campbell River Community Arts Council, Campbell River, British Columbia.

Brasileiro, L., Machado, R. and Aguiar, L. (2022) 'Ecosystems services provided by bats are at risk in Brazil', *Frontiers of Ecological Evolution*, 10, p. 852177. https://doi.org/10.3389/fevo.2022.852177.

Briggs, W. R. (2006) 'Physiology of plant responses to artificial night lighting', in Rich, C. and Longcore, T. (eds) *Ecological consequences of artificial night lighting*. Washington, DC: Island Press, pp. 389–411.

Bruce-White, C. and Shardlow, M. (2011) *A review of the impact of artificial light on invertebrates*. Peterborough: Buglife—The Invertebrate Conservation Trust.

Chinn, M. (2013) 'Sensing the wind: The timely music of nature's memory', *Environmental Philosophy*, 10(1), pp. 25–38.

Cuomo, C. (2002) 'On ecofeminist philosophy', *Ethics and the Environment*, 7(2), pp. 1–11. https://doi.org/10.1353/een.2002.0016.

Davies, T., Bennie, J. and Inger, R. (2013) 'Artificial light pollution: Are shifting spectral signatures changing the balance of species interactions?' *Global Change Biology*, 19, pp. 1417–1423.

Dill, K. (2021) 'In defense of wild night', *Ethics, Policy, and Environment*, 25(2), pp. 153–177. https://doi.org/10.1080/21550085.2021.1904496.

Evans, D., Pocock, M. and Memmott, J. (2013) 'The Robustness of a network of ecological networks to habitat loss', *Ecology Letters*, 16(7), pp. 844–852.

Falcón, J., Torriglia, A., Attia, D., Viénot, F., Gronfier, C., Behar-Cohen, F., Martinsons, C. and Hicks, D. (2020) 'Exposure to artificial light at night and the consequences for flora, fauna, and ecosystems', *Frontiers of Neuroscience*, 14, p. 602796. https://doi.org/10.3389/fnins.2020.602796.

Fox, R. (2013) 'The decline of moths in Great Britain: A review of possible causes', *Insect Conservation and Diversity*, 6, pp. 5–19.

Frank, K. (2006) 'Effects of artificial night lighting on moths', in Rich C. and Longcore, T. (eds.) *Ecological consequences of artificial night lighting*. Washington, DC: Island Press, pp. 305–344.

Frost, R. (2002 [1923]) *The poetry of Robert Frost: The collected poems*. Lathem, E. C. (ed.), New York, NY: Holt Paperbacks.

Fleming, T. (1988) *The short-tailed fruit bat: A study in plant-animal interactions*. Chicago, IL: University of Chicago Press.

Futsaether, C., Vollsnes, A., Kruse, O. M. O., Otterholt, E., Kvaal, K. and Eriksen, A. B. (2009) 'Effects of the Nordic photoperiod on ozone sensitivity and repair in different clover species studied using infrared imaging', *Ambio,* 38, pp. 437–443.

Gaston, K., Davies, T., Nedelec, S. and Holt, L. (2017) 'Impacts of artificial light at night on biological timings', *Annual Review of Ecology, Evolution, and Systematics*, 48(1), pp. 49–68.

Ghanem, S. and Voigt, C. (2012) 'Increasing awareness of ecosystem services provided by bats', in *Advances in the Study of Behavior*, 44, pp. 279–302.

Giavi, S., Fontaine, C. and Knop, E. (2021) 'Impact of artificial light at night on diurnal plant-pollinator interactions', *Nature Communications*, 12, p. 1690. https://doi.org/10.1038/s41467-021-22011-8.

Gilhen-Baker, M., Roviello, V. and Beresford-Kroeger, D. (2022) 'Old growth forests and large old trees as critical organisms connecting ecosystems and human health', *Environmental Chemistry Letters*, 20, pp. 1529–1538. https://doi.org/10.1007/s10311-021-01372-y.

Hölker, F., Wolter, C., Perkin, E. and Tockner, K. (2010) 'Light pollution as a biodiversity threat', *Trends in Ecology and Evolution*, 25(12), pp. 681–682. https://doi.org/10.1016/j.tree.2010.09.007.

Hourdequin, M. and Wong, D. (2005) 'A relational approach to environmental ethics', *Journal of Chinese Philosophy*, 32, pp. 18–33.

Ikei, H., Song, C. and Miyazaki, Y. (2015) 'Physiological effect of olfactory stimulation by Hinoki Cypress (Chamaecyparis obtusa) leaf oil', *Journal of Physiological Anthropology*, 34(1), p. 44. https://doi.org/10.1186/s40101-015-0082-2.

Kaplan, S. (1995) 'The restorative benefits of nature: Toward an integrative framework', *Journal of Environmental Psychology*, 15(3), pp. 169–182. https://doi.org/10.1016/0272-4944(95)90001-2

Kimmerer, R. (2013) *Braiding sweetgrass*. Minneapolis, MN: Milkweed Editions.

Leopold, A. (1949 (1986)) *A sand county almanac (outdoor essays and reflections)*. New York, NY: Ballantine Books.

Lewanzik, D. and Voigt, C. (2014) 'Artificial light puts ecosystem services of frugivorous bats at risk', *Journal of Applied Ecology*, 51, pp. 388–394. Available at: https://besjournals.onlinelibrary.wiley.com/doi/epdf/10.1111/1365-2664.12206 (Accessed: 29 April 2023).

Li, Q., Morimoto, K., Nakadai, A., Inagaki, H., Katsumata, M., Shimizu, T., *et al.* (2007) 'Forest bathing enhances human natural killer activity and expression of anti-cancer proteins', *International Journal of Immunopathology and Pharmacology*, 20, pp. 3–8.

Liu, Z., Lv, Y., Ding, R., Chen, X. and Pu, G. (2020) 'Light pollution changes the toxicological effects of cadmium on microbial community structure and function associated with leaf litter decomposition', *International Journal of Molecular Science*, 21(2), p. 422. https://doi.org/10.3390/ijms21020422.

Lowery, S., Blackman, S. and Abbate, D. (2009) *Urban movement patterns of lesser long-nosed bats (Leptonycteris curasoae): Management implications for the habitat conservation plan within the city of Tucson and the town of Marana*. Phoenix, AZ: Arizona Game and Fish Department.

MacGregor, C., Pocock, M., Fox, R. and Evans, D. (2015) 'Pollination by nocturnal Lepidoptera, and the effects of light pollution: A review', *Ecological Entomology*, 40, pp. 187–198. https://doi.org/10.1111/een.12174.

Margulis (Sagan), L. (1967) 'On the origin of mitosing cells', *Journal of Theoretical Biology*, 14(3), pp. 225–274.

Montoya, J., Pimm, S. and Solé, R. (2006) 'Ecological networks and their fragility', *Nature*, 442, pp. 259–264. https://doi.org/10.1038/nature04927.

Morrison, D. (1978) 'Foraging ecology and energetics of the frugivorous bat Artibeus Jamaicensis', *Ecological Society of America*, 59(4), pp. 716–723.

Muscarella, R. and Fleming, T. (2007) 'The role of frugivorous bats in tropical forest succession', *Biological Reviews*, 82(4), pp. 573–590.

Naess, A. (1973) 'The shallow and the deep, long-range ecology movement. A summary', *Inquiry*, 16, pp. 95–100.

Pu, G., Zeng, D., Mo, L., Liao, J., Chen, X., Qiu, S. and Lv, Y. (2020) 'Artificial light at night alter the impact of arsenic on microbial decomposers and leaf litter decomposition in streams', *Ecotoxicology and Environmental Safety*, 191, p. 110014. https://doi.org/10.1016/j.ecoenv.2019.110014

Sheldrake, M. (2020) *Entangled life*. New York, NY: Random House.

Shiva, V. (1988) *Staying alive: Women, ecology, and survival in India*. London: Zed Books Ltd.

Simard, S. (2021) *Finding the mother tree*. New York, NY: Random House.

Sower, L., Shorey, H. and Gaston, L. (1970) 'Sex pheromones of noctuid moths. XXI. Light-dark cycle regulation and light inhibition of sex pheromone release by females of *Trichoplusia ni*', *Annals of the Entomological Society of America*, 63, pp. 1090–1092.

Stone, E., Jones, G. and Harris, S. (2009) 'Street lighting disturbs commuting bats', *Current Biology*, 14(19), pp. 1123–1127. https://doi.org/10.1016/j.cub.2009.05.058.

Stone, E., Jones, G. and Harris, S. (2012) 'Conserving energy at a cost to biodiversity? Impacts of LED lighting on bats', *Global Exchange Biology*, 18(8), pp. 2458–2465.

Stone, T. (2018) 'Re-envisioning the nocturnal sublime: On the ethics and aesthetics of nighttime light', *Topoi*, 40, pp. 1–11. https://doi.org/10.1007/s11245-018-9562-4.

Velez Ramirez, A., van Ieperen, W., Vreugdenhill, D. and Millenaar, F. (2011) 'Plants under continuous light', *Trends in Plant Science*, 16(6), pp. 310–318. https://doi.org/10.1016/j.tplants.2011.02.003.

Weitz, H., Ballard, A., Campbell, C. and Killham, K. (2001) 'The effect of culture conditions on the mycelial growth and luminescence of naturally bioluminescent fungi', *FEMS Microbiology Letters*, 202(2), pp. 165–170. https://doi.org/10.1111/j.1574-6968.2001.tb10798.x.

Winfree, R., Bartomeus, I. and Cariveau, D. (2011) 'Native pollinators in anthropogenic habitats', *Annual Review of Ecology, Evolution, and Systematics*, 42, pp. 1–22.

10 Darkening cities as urban restoration

Taylor Stone

Restoring darkness

Our nights are getting brighter. A recent multi-year citizen science study has shown this trend to be seriously underestimated. Previous estimates of the rate of annual increase were around 2–3%; it was posited that due to previous measurement techniques combined with the rapid introduction of LED outdoor lighting, the current rate of increase is likely closer to 10% (Kyba *et al.*, 2023). This would seem to reaffirm earlier critiques and observations that the widespread introduction of LEDs could lead to an increase in brightness and energy use, as it offers a more efficient means of producing artificial light and is therefore susceptible to a rebound effect (Gandy, 2017; Kyba, Hänel and Hölker, 2014; Kyba *et al.*, 2017). A commentary on this study argues that we must increasingly see artificial light as "the pollutant it really is" (Falchi and Bará, 2023, p. 235). Such a framing, the commentators assert, can allow for caps and red lines on total usage and illuminance levels.

More needs to be done to increase awareness of this issue and to implement effective policies and strategies for curbing the negative impacts of light pollution. Yet these findings arguably also speak to the limited effectiveness of the concept of light pollution as a problem frame (see Stone, 2017), and the need for new approaches to lighting policy and design. An alternative approach is to position environmental impacts not as a constraint but instead as an opportunity to re-envision the goals driving night-time lighting. Efforts need to be re-imagined from the ground up, with environmental values embedded in policies, lighting strategies, and technical innovations as positive design goals. For this, a robust guiding concept is required to articulate and operationalise the environmental values at stake. This, in turn, must rely on a deeper philosophical reflection into how our lighting technologies relate to their environments. Further, they must embody a narrative that offers new ways of relating to urban darkness and dark skies, showing why these facets of night-time environments should matter to us (see O'Neill, Holland and Light, 2008). Put more simply: alongside the need for quantifiable measurements and resultant policies, we need a new story about urban darkness. For this, I draw inspiration from de-Shalit's (2003) notion of urban restoration, which concerns ideas of the good we want to conserve and foster as much as physical artefacts. In doing so, I also attempt to align two relevant

DOI: 10.4324/9781003408444-14
This chapter has been made available under a CC-BY-NC-ND 4.0 license

contemporary narratives: first, the concerns over the "loss of the night" and resultant dark sky preservation efforts, largely focused on rural or remote areas; second,
movements looking to re-think the role of "nature" within urban environments and
the relationship between emerging technologies and the management of urban ecosystems. We can see this in biophilic design (McDonald and Beatley, 2021) and the
nascent concept of the "internet of nature" (Galle, Nitoslawski and Pilla, 2019).
In this sense, I explore the converge of dark sky preservation with contemporary
visions of urban futures.

Finding a new story requires new ways of thinking about urban darkness and
the features it makes possible, such as dark skies. This requires a shift in perspective away from negative associations with danger, fear, or poverty, and towards
conceiving darkness as a source of positive environmental value (Edensor, 2015,
2017; Stone, 2018). "Dark design" has been proposed as a new design strategy
that strives to reduce the impacts of light pollution, as well as raise awareness
about these challenges (Dunn, 2020). Practical approaches, such as "dark infrastructures" (Narboni, 2017), "dark ecological networks" (Challéat *et al.*, 2021),
and "dark acupuncture" (Stone, Dijkstra and Danielse, 2021), can arguably achieve
these goals through the conscientious design and use of artificial light. These nascent approaches to the design of urban nightscapes not only aim to reduce the
negative effects of light pollution but actively re-imagine the relationship between
(artificial) light and (natural) darkness.

This chapter does not reflect upon or evaluate the merits of specific dark design
strategies. Rather, it is about the narratives these growing strategies devise in order
to address urban darkness and how this can offer a foundational, unifying ideal for
re-designing urban lighting with and for darkness. I begin from the perspective that
the environmental problems we face are cultural, not technical:

> It is not a question of our encountering the crisis and resolving it through
> technology. The crisis is not simply something we can examine and resolve.
> We *are* the environmental crisis. The crisis is a visible manifestation of our
> very being . . . The environmental crisis is inherent in everything we believe
> and do; it is inherent in the context of our lives.
>
> (Evernden, 1993, p. 128)

There is no environment "out there", and likewise there is no "polluting light"
separate from the technologies and infrastructures we design, build, and maintain.
A new evaluative lens is required that offers new perspectives on *how we think
about* urban nights and how and why we choose to illuminate our world after dark.
Only then can we see how technical innovations can be utilised towards responsible ends.

In what follows, I situate darkness and dark skies as a type of urban restoration
and argue that *darkening cities* should be advanced alongside the goal of greening
cities. I position urban restoration holistically, as an attempt to "breathe life into
an urban story that has fallen into decay and that we find meaningful, or valid, or
good" (de-Shalit, 2003, p. 7). For this, we need a robust understanding of darkening

cities that positions it not only as a technical or quantifiable target but as a moral and aesthetic ideal for the future of urban nightscapes. I argue that darkening cities can be best understood as an act of *repairing the night*, which is itself an act of repairing our lighting technologies. I draw from the philosophy of technology, specifically recent work on maintenance and repair, to articulate darkness as an urban feature co-constituted by our lighting technologies and infrastructures. This highlights the contemporary reality that (in developed, urbanised areas of the world) we no longer directly experience "natural darkness" in our daily lives; our urban nights are mediated by artificial lighting (see Verbeek, 2011). Because of this, restoring urban darkness needs to focus on repairing our lighting technologies and more generally asking what material *and* social aspects of this infrastructure we are aiming to repair. This, I assert, concerns the relationship between cities and their ecological sense of place. Understanding repair in this way leads into a discussion of the aesthetics of darkened cities, and the type of experiences and interactions with darkness we should aim to foster. For this, I consider an aesthetics of wildness, with darkness as an act of *rewilding the night* (Dill, 2021). Crucially, these technological and environmental concerns are intertwined—to think about a robust conceptualisation of darkening cities as urban restoration, we need to understand the repair of lighting technologies and the rewilding of urban nights as reinforcing one another. It is *through* our lighting that we repair our nights, and therefore *through* these acts of repair that we can create conditions for wild urban nights. To repair (and darken) our urban nightscapes, we need to bring some wildness into our lighting. Doing so can provide a new story for urban darkness and one that can be informed by creative and innovative approaches to dark design.

Repairing the night

Night is a time of maintenance for cities, when essential acts such as cleaning and the repair of critical infrastructures occurs (Shaw, 2022). But we can also ask if urban nights themselves, and therefore the lighting technologies that co-constitute night-time spaces, are in need of repair. I do not mean literal acts of repairing streetlights that have failed or require replacement (although this analysis can point towards what the goals of these acts should be). Rather, I use repair as a metaphor to conceptualise artificial illumination as a formative technology with regard to how we experience and interact with urban darkness. The 21st century will be one of rapid urbanisation, meaning that the majority of us will spend most of our lives in spaces that no longer have natural night-time conditions or access to dark skies (Falchi *et al.*, 2016). In urban spaces, we have increasing control over the qualities and spatial distribution of darkness, for better and for worse. If we accept this idea—that urban nights are mediated by our technologies—then this act of urban restoration becomes focused on restoring urban darkness *through* our lighting technologies. Doing so can surface a relational facet of urban lighting central to the concept of darkening cities.

Maintenance is typically framed as a practice through which we provide stability to something over time, aligning with an ontology that emphasises the design phase as the moment in which the (largely static) qualities of an artefact are established. A common view of maintenance is thus as constituting acts of conservation (Young, 2021a), in the sense of acting *against* the passage of time—working to combat the inevitable processes of breakdown, erosion, corrosion, decay, malfunction, and obsolescence. The goal is to maintain systems and artefacts in largely the same way as they currently operate. However, an alternative is to adopt a process-oriented view that highlights the change to artefacts and infrastructures such practices facilitate (Young, 2021a). This includes acts of repair as "subtle acts of care by which order and meaning in complex sociotechnical systems are maintained and transformed, human value is preserved and extended, and the complicated work of fitting to the varied circumstances of organizations, systems, and lives is accomplished" (Jackson, 2014, p. 222).

While often focused on the restoration and preservation of existing systems and relations, maintenance and repair can also instantiate processes of change and evolution through the adjustment, tweaking, adaptation or more radical changes to systems and artefacts. This highlights the temporal and dynamic nature of technology as formed by processes that are extended and guided through time. This asks us to appreciate that no technology or system is completely static but is necessarily dynamic and evolving. Through such processes, we see the continuous processes of adaptation and change via external influences and deliberate choices as fundamental to the interaction of artefacts with their world (Young, 2021a, 2021b). This is a useful reminder that all artefacts are dynamic and necessarily interact with the world in some ways and that the results of these interactions often necessitate maintenance and repair—a perspective that points towards the temporal dimensions of wildness discussed in the next section. Deliberate acts of repair, however, need not only be concerned with maintaining the status quo. Henke and Sims (2020) point to a paradoxical feature of maintenance and repair of infrastructure during our era of environmental crises; they contend that "we cannot develop data and theories about 'the world' without the very systems that are creating wide-scale change" (Henke and Sims, 2020, p. 122). To confront our current situation, we need to think about "repairing repair", or what they term reflexive repair. Here reflexivity has two layers: in a causal sense—in that the operation of infrastructures cause impacts that (re-)shape their own operating environment—and in the sense of self-aware thinking—of those designing and managing the systems to appreciate potentially negative feedback loops.

Urban lighting infrastructure is no exception; the very concept of light pollution carries these layers of reflexivity. We are now confronted with the problem of having *too much light* (or at least too much poorly designed lighting). But in searching for a way forward, we quickly encounter a puzzle. A precondition for the concept of light pollution is an abundance of artificial lighting. Urban lighting has played such a defining role in shaping its own operating environment, namely darkness, that we are now forced to confront the negative feedback loops caused by these impacts. The shift towards desiring and seeking to preserve darkness is—at least in part—a result of the proliferation of night-time illumination. Paradoxically, the contemporary desire for more darkness actively de-values the very technology that necessitates its own moral re-evaluation. Accordingly, a process-oriented view of technology suggests that we should abandon completeness as a goal for any framework for mitigating light pollution or creating darker cities. Any successful re-introduction of urban darkness, as well as light pollution mitigation, will necessarily change the relative meaning and importance of associated values and goals for urban nightscapes. This should not be seen as a problem but illustrative of the fact that urban darkness is not something that can be "solved" with any finality. (Natural) darkness and (artificial) light are competing but also complementary—there will be a continual striving for an ideal balance, which will itself evolve as a result of urban lighting policies and strategies (Stone, 2021b).

This highlights an important aspect of repair practices, namely, that such acts have both a material and social function. Repairing an artefact or system not only is an act of restoring functionality but often serves to repair social relations and values (Henke and Sims, 2020). This is critical for how we conceptualize restoring urban darkness via repairing urban lighting. Materially, darkening cities can serve to repair, restore, and preserve urban ecologies. This requires positioning dark skies as a type of "natural infrastructure" (McDonald and Beatley, 2021)—like other green and blue aspects of cities, such as urban forests—that can reduce energy consumption while improving biodiversity and the wellbeing of citizens. Yet dark skies also carry an interrelated social dimension, offering the possibility of restoring an ecological and cosmological sense of place, in the process disrupting the geographical dualism between *built* and *natural* environments (Stone, 2021a). Importantly, this makes repair more than simply acts of repairing an infrastructure or even of restoring nocturnal ecosystems; they are also repairing our *relation* to urban darkness and the night sky.

Rewilding the night

What conditions or qualities of night-time environments should we strive to achieve via the repair of lighting technologies? If lighting technologies were to embody a reflexive and relational ethos, it would point towards a mode of thinking (and subsequent design interventions) that moves away from binary categories. Instead of restoring the "natural night sky" as a static objective and one in opposition to a polluted night sky, we could embrace the rhythms and temporality of darkness. This could bring some wildness back into our cities.

Rewilding is categorised as a specific form of ecological restoration, with its own set of conservation strategies as well as ethical and aesthetic issues. Prior and Brady (2017, p. 34) define rewilding as "a process of (re)introducing or restoring wild organisms and/or ecological processes to ecosystems where such organisms and processes are either missing or are 'dysfunctional'". However, it is important to note that there is no consensus about the exact definition and scope of rewilding. Because of this, it is better to understand it as a *cluster concept*: there are multiple characteristics that count towards something falling under the concept, and while some of these must be satisfied, none are necessary conditions (Gammon, 2018). Importantly, it is not about restoring a specific state but rather about creating conditions for natural processes to evolve and, ideally, flourish.

A few points regarding the characteristics of rewilding are important to empha-sise, as they have ramifications for the conceptualisation of *urban* rewilding and the role that urban lighting can play in facilitating these processes. First, wildness should not be conflated with wilderness. Wildness can be a component *of* wilder-ness, but they are not the same thing. While wilderness has a (discrete) spatial dimension, wildness is concerned with processes and the (eventual) autonomy of the more-than-human world. Because of this, wildness can arguably be found in cultural landscapes, and at all scales, from the starry night sky to weeds in sidewalk cracks to ants in your backyard. This means that rewilding—unlike other forms of ecological restoration—does not require the erasure or removal of humans and cultural landscapes. Thus, rewilding should be understood as relational rather than binary, implementable on a variety of scales and intensities, and potentially emer-gent in non-homogenous ways across a landscape (Prior and Brady, 2017).

Efforts to rewild are often targeted at remote or rural areas, with the intention of removing negative human influence and allowing for the increasing autonomy of wildlife and natural processes. However, there is also an increasing interest in *urban* rewilding (Owens and Wolch, 2019), which I emphasise as a cluster concept, not a specific strategy. Urban rewilding, then, is not a discrete criterion to be satisfied or achieved but an overarching principle to orient urban development. As a form of *urban* ecological restoration, it necessarily features characteristics that are unique to its context and not found in remote areas. First, it often works with(in) the remains of technical infrastructures, such as vacant lots or post-industrial sites. Second, it often combines biodiversity conservation with cultural ecosystem services. Third, and importantly, urban rewilding often requires some (in)direct human intervention (such as management and regulation) to maintain (Kowarik, 2013, 2018). There are a growing number of promising urban rewilding projects around the world. For a taxonomy of urban wildness and examples from the German context, see Stöcker, Suntken and Wissel (2014), with other examples provided by Chicago's waterfront (Northerly Island) and North Branch Canal (the Wild Mile project).

Acts of rewilding can be positioned as conservation strategies, for example, by removing artificial illumination from specific areas to allow certain species to carry out their previously disrupted nocturnal activities. Yet to be successful, rewilding must be seen as a political project and become meaningful in our lives: "Rewild-ing in practice has to navigate between the ecological facts that abound in the scientific literature, and the ways in which these can become meaningful in actual human lives" (Tanasescu, 2017, p. 345). Arguably, political facets become even more pressing in any attempts to foster wildness within urbanised spaces. Rewild-ing cities is not only about restoring urban ecosystems but also asking what forms wildness should take in the design and management of urban technologies and in the daily lives of urban dwellers.

This opens up complex questions about what a rewilded urban night can and should entail, notably regarding the relationship between urbanism and wildness, and the intentions and presence of humans in urban rewilding efforts. Rewilding as defined earlier is largely about embracing processes rather than idealised end states, as well as the uncertainty and unpredictability that this entails. Cities are by definition anthropocentric in design and use; they are built by humans, for humans. They seem to be the antithesis of wilderness. This can give rise to the critique that they cannot be (re)wilded at all. However, as discussed, rewilding can take a variety of scales and degrees—it need not be an all or nothing categorisation. Further, rather than take this as a set of conflicting or contrasting ideals, we can instead ask how this forces us to confront our presuppositions underlying the dualism between built and natural environments. An urban focus can arguably help with the process of *human* rewilding (Maffey and Arts, 2023), for many ecological rewilding efforts, while laudable, effectively treat the symptom and not the cause. Humans need to be included in these processes—not just as stakeholders but engaged within and requiring some rewilding themselves. This can allow us to confront the dichotomies of human-nature as well as built-natural.

Vogel (2015) reminds us that *re*-wilding is in fact a misnomer of sorts because cities (and the artefacts that make them up) are already in many ways wild. Cities are embedded within and rely upon natural systems, and artefacts of all scales often have unexpected and unintended consequences. This means that urban rewilding is less about changing course and more about ceasing to fight against the current. As Vogel (2015, p. 110). states,

> To call [an artefact] wild would be to say that in it, forces are currently operating independently of humans, although it will have been human action (and human purpose) in the first place that led to these forces operating. To produce a wild artifact might mean to (intentionally) put natural forces into action and then (intentionally) to *let them go*, in ways that are fundamentally unpredictable and outside one's control'

He continues (ibid., p. 111):

> Once we abandon the fetish that only a landscape that humans have never touched could possibly be wild, we might begin to see that even ongoing human action within a landscape could be consistent with its ongoing wildness, and could indeed help to maintain it. And once we see that a landscape is a dynamic entity, always undergoing transformation through the actions of the various organisms and weather and geological forces that form it, we might come to realize that the point of restoration is not the reproduction of a particular *thing* but rather the putting into play of processes—of wildness— that we then allow to operate unpredictably and unimaginably in ways that are outside our ability to control. To recognize this would in turn be to see that the wildness that we're after is there all the time, throughout the restoration process; it's not something that comes in at the end, something we *produce*, but rather something that we *use*.

In this sense, rewilding the night is not striving for a space free of human influence but rather re-orienting our infrastructures to align with their ecological and geographical context and the natural rhythms therein.

Rewilding will bring to the fore the temporal dimensions of cities and their nightscapes, as a normative and aesthetic consideration (Lehtinen, 2021). At a practical level, systems such as smart lighting would likely feature dynamic programs responsive to diurnal and seasonal cycles, weather, and changing wildlife needs. However, there will also need to be a consideration of the long-term, intergenerational dynamics of rewilding on urban form (and vice versa). Rewilding is about fostering relations that will develop and mature over long time spans and be in continuous flux. Because of this, there is no optimal "end" to rewilding the night but rather a set of dynamic and evolving relations. Yet this could align with an evolutionary perspective that positions cities as four-dimensional object co-located in space in time. In this sense, cities are best understood as dynamic and emergent complex systems that are never complete but always in a state of change (Batty, 2018; Bettencourt, 2014; Lehtinen, 2021; Stone, 2021b; Varzi, 2021). That we could see both cities *and* their ecologies as processes rather than stable entities may serve to re-orient urban planning and the role of technologies therein.

Reimagining cities after dark

We can now return to the original motivation: of developing a robust conceptualisation of darkening cities as a type of urban restoration and one to help establish visions for our urban night-time futures. The analysis here highlights the entwinement of the technological and environmental dimensions of urban nightscapes and the transformative possibilities embodied in how we choose to design and use urban lighting. It also reminds us to look beyond reductive or binary thinking regarding the re-introduction of urban darkness. It is often stated that darkness, and features such as the night sky, can be achieved "with the flip of a switch". This is admittedly true. Unlike other objects of ecological restoration, we have not destroyed the starry night sky, just cut ourselves off from this experience. But as our discussion here elucidates, darkening cities should not be solely about turning off lights. There are, of course, common-sense solutions regarding dimming and motion sensors, proper shielding, and regulations to ensure certain types of light are restricted in their use—such as the policies advocated by DarkSky (2023). Such policies make economic and ecological sense and should be a baseline for any city's lighting plans.

However, the notion of darkening cities developed here is not about achieving (near-)natural night-time conditions in urban centres that equates natural conditions with the absence of artificial light. This is neither realistic nor constructive. It would be like assuming that greening cities is about destroying all roads and buildings to bring back an old-growth forest. Rather, it is about navigating the complex relationship between our built and natural environments, towards asking what exactly it means to foster wildness *through* our lighting choices. Elsewhere I have reflected upon Thierry Cohen's *Darkened Cities* series to articulate what a new version of the sublime might be for urban nights and where on the spectrum between our current cities and Cohen's alternative we wish to land (Stone, 2021a). Cohen's series provides dramatic visions of major urban cities with a starry night sky above, reminding us that our city nights *could* look otherwise. However, these images have also been edited to completely remove artificial light and human presence. It is also worth imagining what a *darkened cities* image would look like with a vibrant, inhabited city below the night sky—a space that does not completely erase the aesthetics or function of urban lighting. It would certainly be darker, although not featuring a pristine and unpolluted night sky. However, it would offer a re-oriented urban nightscape and a re-imagining of urban lighting and the possibilities for embracing darkness therein.

We can already see the seeds of such ideas within lighting design. There are companies aiming to commercialise bioluminescent lighting, such as GLOWEE (n.d.), and experimental projects such as the Academy for City Astronauts have proposed visions and strategies aimed at realising a "dark city" (Academie voor Stadsastronauten n.d.). In 2017, the Professional Lighting Design Convention (PLDC) hosted a Design Ideas Competition on "The Future of Urban Lighting" (PLDC, 2017). It featured concepts from eight well-established lighting design firms as well as ten competition entries from design teams. Contributors to the exhibition were asked to imagine how cities will be illuminated in 2053 and to touch upon topics such as luminous materials, bioluminescence, pedestrian ambiance, biorhythms and wellbeing, and personalised lights. The resulting concepts and visions are diverse and inspiring (and not necessarily practical or even desirable) but share a common feature: they do not imagine *brighter* nights but balanced and inviting atmospheres that re-enchant urban lighting *through* the re-introduction of darkness. These projects exemplify the sorts of explorations that can translate the goal of darkening cities into tangible visions and eventually inform the realisation of practical dark design strategies.

Darkening cities is about instilling *lighting technologies themselves* with an ecological and cosmological sense of place—as mediators of the night-time experience of our cities. This will not necessarily re-introduce features such as the night sky in any absolute sense but rather a night-time environment attuned to place and more-than-human needs. In urban spaces, wildness does not merely concern removing technology; it *is* the technology, or at least how it interacts with its environment. It is about using lighting design to rediscover and realign cities with natural rhythms (Eklöf, 2022; Griffiths and Dunn, 2020). Dill (2021) describes these sorts of design strategies as acts of synergistic rewilding, referring to human behaviours that contribute to flourishing ecosystems. Doing so can arguably carry with it a material and social function, in addition to any instrumental benefits such as energy reductions or benefits to biodiversity. Further, it instantiates a process of change in how we view cities after dark and how we choose to intervene through technological means. A fully "dark city" will never be fully achieved; it will be relational and process-based, not a static or absolute end goal. It should not be itemised into a checklist or instrumentalised in the ways that sustainably too often is. Instead, it can serve as a catalyst for new ways of thinking about urban nights. Darkening cities as a foundational goal—and practical dark design strategies that strive to realise this goal—is not just about reducing artificial illumination to acceptable polluting levels. It concerns the fostering of wildness *within* our lighting technologies. In doing so, we can re-align our cities with natural cycles of light and dark.

Conclusion: inventing urban darkness

McDonald and Beatley (2021) share an anecdote about how the Dutch "invented" street trees. Of course, they do not mean to claim that the Dutch somehow invented trees or streets. Rather, they came up with the innovation to line streets with trees to stabilise the canal banks that run through city centres. From this was spurred a new typology of vegetation in cities that persists for a variety of ecological, health, and aesthetic reasons that extend beyond their original purpose. Sharpe (2008, p. 24) makes a similar observation about lighting: "The arrival of artificial light had, almost paradoxically, 'invented' natural light, for no such conceptual category existed before the new technologies posed alternative forms of illumination". The coming years will see profound changes to artificial lighting through the continued and rapid adoption of LEDs, combined with the implementation of smart lighting systems aided by sensors and artificial intelligence. The introduction of novel forms of illumination, such as bioluminescence, may offer new design possibilities, and seemingly disparate technologies, such as autonomous urban robots and vehicles, may disrupt or make redundant the very need for streetlights as they currently exist. With such changes at on the horizon, we will inevitably be confronted with choices about our urban nights. What is now required is a reflexive approach to the technologies that will drive this change, alongside a re-evaluation of the values and goals we strive to realise. Perhaps what we now need is to "invent" a new conception of urban dark skies. This category was not required before but has been created through the attention and advocacy for "natural" dark skies. Instead of seeing urban darkness as dangerous and unsafe, there is an opportunity to position this conceptual category as a tool to re-imagine the future of urban nights. Doing so can offer a novel form of urban restoration and give new meaning to a facet of urban wildness that has been neglected for too long. What story will our lighting tell us about urban darkness?

References

Academie voor Stadsastronauten. (n.d.) *Nachtlab: De donkere stad*. Available at: https://stadsastronaut.nl/ (Accessed: 1 April 2021).

Batty, M. (2018) *Inventing Future Cities*. Cambridge, MA: The MIT Press.

Bettencourt, L. M. A. (2014) 'The uses of big data in cities', *Big Data*, 2(1), pp. 12–22. https://doi.org/10.1089/big.2013.0042.

Challéat, S., Barré, K., Laforge, A., Lapostolle, D., Franchomme, M., Sirami, C., Le Viol, I., Milian, J. and Kerbiriou, C. (2021) 'Grasping darkness: The dark ecological network as a social-ecological framework to limit the impacts of light pollution on biodiversity', *Ecology and Society*, 26(1), p. 15. https://doi.org/10.5751/ES-12156-260115.

DarkSky. (2023) *DarkSky*. Available at: http://darksky.org/ (Accessed: 1 August 2023).

de-Shalit, A. (2003) 'Philosophy gone urban: Reflections on urban restoration', *Journal of Social Philosophy*, 34(1), pp. 6–27. https://doi.org/10.1111/1467-9833.t01-1-00162.

Dill, K. M. (2021) 'In defense of wild night', *Ethics, Policy & Environment*, pp. 1–25. https://doi.org/10.1080/21550085.2021.1904496.

Dunn, N. (2020) 'Dark design: A new framework for advocacy and creativity for the nocturnal commons', *The International Journal of Design in Society*, 14(4), pp. 19–30.

Edensor, T. (2015) 'The gloomy city: Rethinking the relationship between light and dark', *Urban Studies*, 52(3), pp. 422–438. https://doi.org/10.1177/0042098013504009.

Edensor, T. (2017) *From light to dark: Daylight, illumination, and gloom.* Minneapolis, MA: University of Minnesota Press.

Eklöf, J. (2022) *The darkness manifesto: How light pollution threatens the ancient rhythms of life.* London: The Bodley Head.

Evernden, L. L. N. (1993) *The natural alien: Humankind and environment.* Toronto, ON: University of Toronto Press.

Falchi, F. and Bará, S. (2023) 'Light pollution is skyrocketing', *Science*, 379(6629), pp. 234–235. https://doi.org/10.1126/science.adf4952.

Falchi, F., Cinzano, P., Duriscoe, D., Kyba, C. C. M., Elvidge, C. D., Baugh, K., Portnov, B. A., Rybnikova, N. A. and Furgoni, R. (2016) 'The new world atlas of artificial night sky brightness', *Science Advances*, 2(6), e1600377. https://doi.org/10.1126/sciadv.1600377.

Galle, N. J., Nitoslawski, S. A. and Pilla, F. (2019) 'The internet of nature: How taking nature online can shape urban ecosystems', *The Anthropocene Review*, 6(3), pp. 279–287. https://doi.org/10.1177/2053019619877103.

Gammon, A. R. (2018) 'The many meanings of rewilding: An introduction and the case for a broad conceptualisation', *Environmental Values*, 27(4), pp. 331–350. https://doi.org/10.3197/096327118X15251686827705.

Gandy, M. (2017) 'Negative luminescence', *Annals of the American Association of Geographers*, 107(5), pp. 1090–1107. https://doi.org/10.1080/24694452.2017.1308767.

GLOWEE. (n.d.) *Bioluminescence.* Available at: www.glowee.eu/ (Accessed: 15 April 2023).

Griffiths, R. and Dunn, N. (2020) 'More-than-human Nights:Intersecting lived experience and diurnal rhythms in the nocturnal city' in Garcia-Ruiz, M. and Nofre, J. (eds.) *ICNS proceedings.* Lisbon: ISCTE, pp. 203–220. https://eprints.lancs.ac.uk/id/eprint/148807/.

Henke, C. R. and Sims, B. (2020) *Repairing infrastructures: The maintenance of materiality and power.* Cambridge, MA: The MIT Press. https://doi.org/10.7551/mitpress/11771.001.0001.

Jackson, S. J. (2014) 'Rethinking repair', in Gillespie, T., Boczkowski, P. J. and Foot, K. A. (eds.) *Media technologies: Essays on communication, materiality and society.* Cambridge, MA: The MIT Press, pp. 221–239.

Kowarik, I. (2013) 'Cities and wilderness. A new perspective', *International Journal of Wilderness*, 19, pp. 32–36.

Kowarik, I. (2018) 'Urban wilderness: Supply, demand, and access', *Urban Forestry & Urban Greening*, 29(3), pp. 36–347. https://doi.org/10.1016/j.ufug.2017.05.017.

Kyba, C. C. M., Altıntaş, Y. Ö., Walker, C. E. and Newhouse, M. (2023) 'Citizen scientists report global rapid reductions in the visibility of stars from 2011 to 2022', *Science*, 379(6629), pp. 265–268. https://doi.org/10.1126/science.abq7781.

Kyba, C. C. M., Hänel, A. and Hölker, F. (2014) 'Redefining efficiency for outdoor lighting', *Energy & Environmental Science*, 7(6), pp. 1806–1809. https://doi.org/10.1039/C4EE00566J.

Kyba, C. C. M., Kuester, T., Sánchez de Miguel, A., Baugh, K., Jechow, A., Hölker, F., Bennie, J., Elvidge, C. D., Gaston, K. J. and Guanter, L. (2017) 'Artificially lit surface of Earth at night increasing in radiance and extent', *Science Advances*, 3(11), e1701528. https://doi.org/10.1126/sciadv.1701528.

Lehtinen, S. (2021) 'Another look at the city. Emphasizing temporality in urban aesthetics', in Kvokačka, A. and Giombini, L. (eds.) *Everydayness. Contemporary aesthetic approaches.* Presov: University of Presov, pp. 31–42.

Maffey, G. and Arts, K. (2023) 'Human rewilding: Practical pointers to address a root cause of global environmental crises', in Hawkins, S., Convery, I., Carver, S. and Beyers, R. (eds.) *Routledge handbook of rewilding*. London: Routledge, pp. 374–382.

McDonald, R. and Beatley, T. (2021) *Biophilic cities for an urban century*. London: Palgrave Macmillan.

Narboni, R. (2017) 'Imagining the future of the city at night', *Architect*. Available at: www. architectmagazine.com/technology/lighting/imagining-the-future-of-the-city-at-night_o.

O'Neill, J., Holland, A. and Light, A. (2008) *Environmental values*. New York, NY: Routledge.

Owens, M. and Wolch, J. (2019) 'Rewilding cities', in Pettorelli, N., Durant, S. M. and du Toit, J. T. (eds.) *Rewilding*. Cambridge: Cambridge University Press, pp. 280–302. https://doi.org/10.1017/9781108560962.014.

PLDC. (2017) *'The future of urban lighting'—the visions of eight masters of lighting design (results of the design ideas competition 2017)*. Gütersloh: VIA-Verlag, Joachim Ritter e.K.

Prior, J. and Brady, E. (2017) 'Environmental aesthetics and rewilding', *Environmental Values*, 26(1), pp. 31–51. https://doi.org/10.3197/096327117X14809634978519.

Sharpe, W. C. (2008) *New York nocturne: The city after dark in literature, painting, and photography*. Princeton, NJ: Princeton University Press.

Shaw, R. (2022) 'Geographies of night work', *Progress in Human Geography*, 46(5), pp. 1149–1164. https://doi.org/10.1177/03091325221107638.

Stöcker, U., Suntken, S. and Wissel, S. (2014) *A new relationship between city and wilderness. A case for wilder urban nature*. Berlin: Deutsche Umwelthilfe.

Stone, T. (2017) 'Light pollution: A case study in framing an environmental problem', *Ethics, Policy and Environment*, 20(3), pp. 279–293. https://doi.org/10.1080/21550085.201 7.1374010.

Stone, T. (2018) 'The value of darkness: A moral framework for urban nighttime lighting', *Science and Engineering Ethics*, 24(2), pp. 607–628. https://doi.org/10.1007/ s11948-017-9924-0.

Stone, T. (2021a) 'Re-envisioning the nocturnal sublime: On the ethics and aesthetics of nighttime lighting', *Topoi*, 40(2), pp. 481–491. https://doi.org/10.1007/s11245-018-9562-4.

Stone, T. (2021b) 'Design for values and the city', *Journal of Responsible Innovation*, 8(3), pp. 364–381. https://doi.org/10.1080/23299460.2021.1909813.

Stone, T., Dijkstra, I. and Danielse, T. (2021) 'Dark acupuncture: A design strategy for sustainable lighting', *International Journal of Sustainable Lighting*, 23(2), pp. 70–87. https:// doi.org/10.26607/ijsl.v23i2.112.

Tanasescu, M. (2017) 'Field notes on the meaning of rewilding', *Ethics, Policy and Environment*, 20(3), pp. 333–349. https://doi.org/10.1080/21550085.2017.1374053.

Varzi, A. C. (2021) 'What is a city?' *Topoi*, 40(2), pp. 399–408. https://doi.org/10.1007/ s11245-019-09647-4.

Verbeek, P.-P. (2011) *Moralizing technology: Understanding and designing the morality of things*. Chicago, IL: University of Chicago Press.

Vogel, S. (2015) *Thinking like a mall: Philosophy after the end of nature*. Cambridge, MA: The MIT Press.

Young, M. T. (2021a) 'Maintenance', in Michelfelder, D. and Doorn, N. (eds.) *The Routledge handbook of the philosophy of engineering*. London: Routledge, pp. 356–368.

Young, M. T. (2021b) 'Now you see it (now you don't): Users, maintainers and the invisibility of infrastructure', in Nagenborg, M., Stone, T., González Woge, M. and Vermaas, P. E. (eds.) *Technology and the city: Towards a philosophy of urban technologies*. Cham: Springer International Publishing, pp. 101–119. https://doi.org/10.1007/978-3-030-52313-8_6.

Part 5

Dark sky communities

11 Designing with the dark

Kerem Asfuroglu

Introduction

This chapter draws on the author's professional experience of working in the light-ing industry for more than a decade. Mindful of the shifting purpose and cultural meanings of artificial light that have contributed to the problems of over-illumination and consequent negative environmental, social, and health impacts, I seek to ques-tion how alternative perspectives and practices might be deployed. I was a junior lighting designer at the start of my career when I watched the documentary, *The City Dark* (2011). It is a powerful film which opened my eyes to this issue. I felt a profound connection and responsibility towards this matter because I dealt with light as a profession. Accordingly, I have sought to develop practical precedents in which the judicious and considerate use of light is applied in the context of work-ing with committed, inventive communities who have striven to restore and protect their dark skies. In exploring the different attributes of three selective projects, I exemplify potential ways in which we can beneficially design with and for dark sky spaces and the people who live within them.

As many chapters in this book confirm, the experience of a dark sky remains exhilarating for many, sparking imaginations, shifting perspectives, and transform-ing the aesthetic experience of the night (Figure 11.1). Moreover, once an aware-ness of the values and qualities associated with darkness is stirred, it can be difficult to unsee how light pollution denies a richer experience of the night. Such senti-ments can subsequently pool to further foster collective strategies amongst kindred minds that can empower movement towards the protection of dark skies.

However, as is also evident in contemporary times, darkness continues to be associated with a host of negative values (Ekirch, 2005; Koslofsky, 2011). To coun-ter these malign connotations, I have named my studio *Dark Source* to underline my commitment to improving the positive perception of darkness. Yet in celebrating and valuing darkness, it is essential to grasp how light and dark do not imply their most extreme qualities of blinding radiance and utter blackness since their shifting intensities and levels allow us to visually discern the complex variety of the world. Indeed, the natural night is rarely pitch black but is a diverse and dynamic realm enriched with stars, moon, and weather events whose potency can be enhanced by avoiding excessive illumination (Commission for Dark Skies, 2021).

DOI: 10.4324/9781003408444-16
This chapter has been made available under a CC-BY-NC-ND 4.0 license

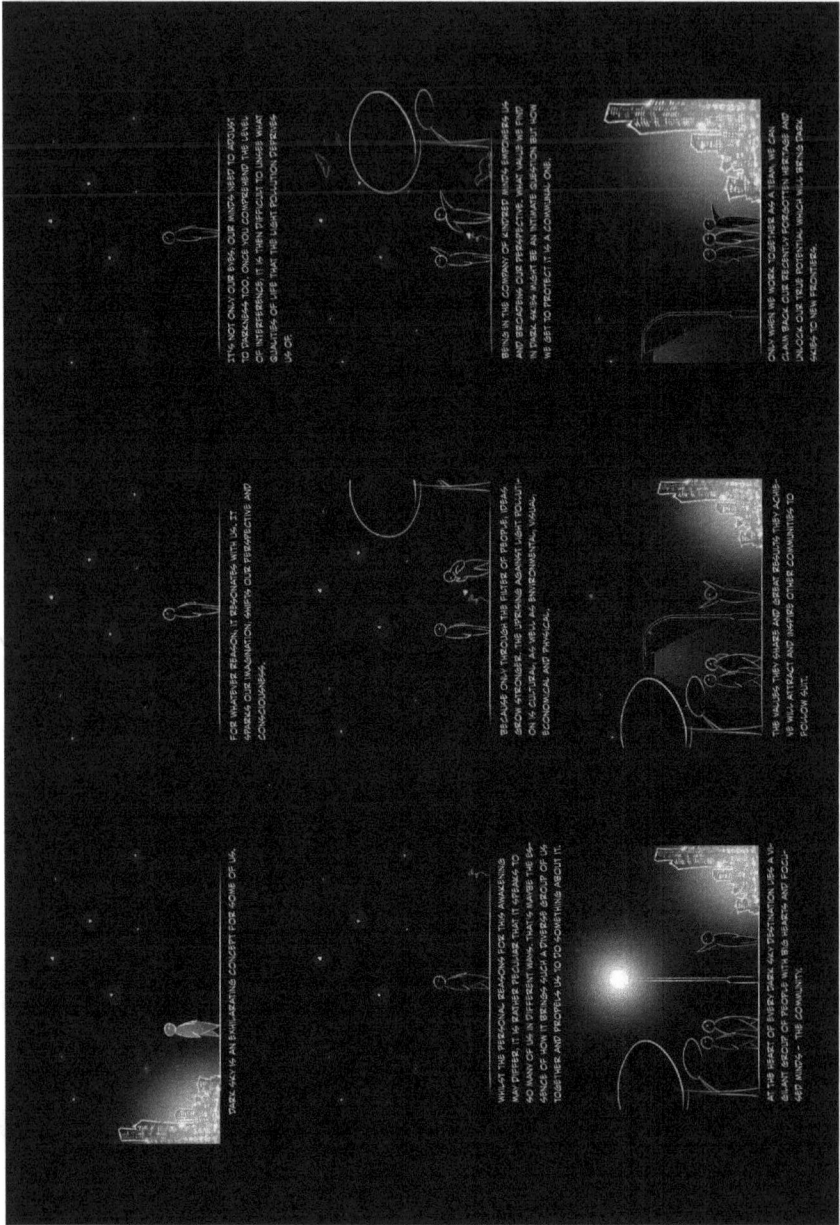

Figure 11.1 Dark Skies Community from Dark Source Graphic Stories at *Arc Magazine*.

Every project I have worked on and every community who have been engaged in lighting projects have been different, with diverse strengths, weaknesses, aspirations, and goals. Similarly, meanings and feelings about light and dark vary between individuals and groups, and these inform the design process. Yet once established, lighting schemes in support of dark skies require committed parties to cherish and nurture them to ensure their survival. This is especially so in an age in which we have never produced and consumed more extensive illumination (Crary, 2013). Initiating cultural change towards dark skies constitutes a continuous struggle between stubborn negative historical notions about darkness and future understandings and values that regard darkness more positively.

Now that the significance of the dark skies movement has claimed more mainstream attention, various mega-projects and resorts are seeking to capitalise on eco-tourism for marketing purposes. While such large-scale projects undoubtedly mark a progressive development, one disadvantage is the notion that dark skies projects can be copied and replicated in any setting. Such an assumption fails to consider the grassroots, bottom-up nature of this movement, qualities that are essential to its longevity. I contend that the role of the small communities who seek to support the reign of dark skies in their own environments are all too often overlooked. It is these smaller-scale, community-oriented projects with whom I have worked and with which I have been most concerned.

Every design project I have worked on is a collaboration. It relies on listening and understanding those people with whom you work and subjugating your own ego and visual ambition. I argue that a project should not be solely concerned with fulfilling the individual vision of the light designer but should develop from the intersection of user demand, environmental needs, and budgetary constraints. The unhindered appetite for visual embellishment without consultation has already caused enough damage to the environment through the generation of light pollution. In the quest to instal illumination that respects the night, we must make the process of design development inclusive, with the participation of stakeholders sought throughout the project, not merely after it has been finalised (Dark Source & Friends of the Lake District, 2023). In exemplifying these insights and to identify the kinds of lighting approaches that have informed the creation of schemes sympathetic to dark skies goals, I now discuss three different projects and the communities with whom I have worked.

Presteigne, Powys, Wales/UK

My first dark skies project was undertaken as part of the Presteigne and Norton Dark Sky Masterplan, with an application due to be submitted to Dark Sky (formerly the International Dark-Sky Association) to gain recognition as the first Dark Sky Community in Wales, supplementing already existing Welsh dark sky reserves and parks. In contrast to other areas of the UK, 18% of Wales has been assigned dark sky status. In developing environmentally friendly lighting for the residents of the town, a judicious approach has been required in focusing upon street lighting. In 2019, on my first visit, orange-tinted sodium lighting, with its inefficient energy expenditure and lack of longevity, had been addressed by their replacement with LEDs for over a decade, a pioneering initiative that saved Powys County Council a significant amount of money by reducing running costs. However, the largely cool white LEDs introduced more significant light pollution problems, flooding the streets with excessive radiance.

Resistance to this scheme was kindled by a small group of amateur astronomers within the community and I assisted in presenting the potential of a redesigned street-lighting plan to the town council. During our initial meeting, we focused on three key points: to reduce energy waste by installing more advanced LED lighting and control technology, to develop the promotion of the town as a destination renowned for its protection of the environment, and to aesthetically enhance the night-time character and visual experience of the town for residents and visitors. Fortunately, members of the Presteigne and Norton Town Council shared the values articulated in these three core goals and became the greatest advocates of the project, which led to its success.

The greatest strength of this project concerns the ways in which it has engaged, mobilised, and included the wider community. A significant degree of trial and error has been involved with, for instance, efforts at fundraising and compiling unsuccessful grant applications. Yet through the process of developing the project, we were able to attract a great deal of attention and support. Moreover, at the point that we thought that we had exhausted every potential avenue and there was no possible way for this project to be realised, Powys County Council came to the rescue. It became apparent that the street lighting was due for a refurbishment and because they were impressed with the amount of dedication which the town council and the community had displayed, it was agreed that the future refurbishment was going to be carried out with a dark-sky-friendly ethos. This graduated the project from a small community effort to a council-backed initiative.

As a London resident, I have been surprised that in many cases where lighting improvements are planned, little stakeholder participation is sought. By contrast, one of the most successful aspects of the Presteigne project was the organisation of lighting tests on two streets in the town centre. Facilitated by the county and town councils, these comprehensive tests gave the community enough time to experience the proposed lighting and consider its virtues before it was initiated. Three light columns along a 150 m stretch of the High Street, in addition to two heritage light columns on Broad Street, were fitted with diverse kinds of luminaires that possessed a variety of colour temperatures and beam distributions. The residents were invited to provide their opinions, with these new lighting applications kept in place for a few months to provide enough time for the changes to be digested. This granted a sense of ownership by allowing them to participate in the final say, promoting the confidence of both community and local authority.

Subsequently, 380 lighting columns were replaced with warm white LEDs, a change in colour temperature that is critical since it scatters in the atmosphere at a lower rate than cooler forms of illumination. Moreover, the installation of cut-off angles has prevented the light from travelling upwards (Figure 11.2). In addition, cost-efficient ways to enable wireless control and dimming has been initiated

Figure 11.2   Images show the impact of replacing the white cool LEDs with warm white
          bulbs in Presteigne, Powys, Wales. Photo by Leigh Harling Bowen.

through a device known as a driver so that the lights can be grouped and individually programmed according to a range of different scene-settings. Forty per cent of
the lights have been programmed to switch off, while the remaining 60% have been
pre-set to dim in intensity by 50% after midnight. In employing dimming profiles
such as this, light intensity and energy usage are reduced while the longevity of the
luminaires is extended. In this context, where limited social activity takes place in
the town late at night, it is wasteful to keep all lights beaming at full intensity. The
plan has significantly reduced annual carbon dioxide emissions by 4.5 tonnes, and
this has been facilitated by the extension of the scheme into a nearby industrial

estate under the jurisdiction of the Welsh government so that the lighting character is consistent across the town.

The success of this lighting project has prompted Powys County Council to consider rolling similar schemes out across the county, thereby empowering many other communities to pursue dark skies accreditation if they so choose. The Presteigne dark skies plan exemplifies how a small community might be bestowed with an agency and platform to transform their nocturnal environment, yet strikingly, it has had a far more extensive impact across Wales, an outcome that was entirely unanticipated. The scheme has been greatly encouraged by the support of a very proactive local-authority street-lighting department, which is not always the case. In addition, a key dimension of this project has been the time required to enable its successful installation, a lengthy period during which community support and enthusiasm, and local authority convictions and backing have been solicited and grown.

Plas y Brenin, Snowdonia, Wales/UK

The second example I discuss is Plas y Brenin, a centre for outdoor sports located in Snowdonia, North Wales, assigned as a dark sky reserve since 2015, which had suffered from a significant level of light pollution that was visible for miles, impacting on surrounding environmental and aesthetic qualities. This stirred the National Park Authority, in collaboration with Prosiect Nos, the North Wales Dark Sky Partnership, which aims to create a vast area of protected dark skies, to commission Dark Source to develop a lighting design to restore the dark skies. The aim was to install dark-sky-friendly lighting to create an inviting ambiance and experience without condemning people to move and rest within conditions of pitch-black darkness.

Prosiect Nos has developed a unique approach in tackling light pollution. They raise funds and target the light polluters in the area. They bring site-specific dark-sky-friendly lighting solutions to these potential suitors, making it amenable for them to implement the proposals. This partnership is an effective and practical method which achieves quick and eye-catching results. The Plas y Brenin project foregrounded an environmental ethos that aimed to reduce energy waste and impact on biodiversity while exploring further opportunities to enhance the night-time experience. Yet as with many other cases, budgets available for dark skies projects are typically modest, encouraging designers to push the boundaries of creativity with regard to the limited financial means available. Initially envisaged as a car park lighting improvement project, the project expanded into a more holistic scheme that covered the whole site. After investigating the relative costs amongst the diverse range of luminaires offered by international suppliers, budget decisions were made that sought the best value for money.

The existing lighting mostly consisted of fluorescent bulkheads and LED floodlights, illumination that reduced the site's legibility, with glare and uncontrolled light dispersal across space shaping visual experience. Moreover, the lighting was unevenly distributed, with certain areas over-lit and other regions of the site deprived of light. Accordingly, the holistic approach that was devised sought to striking a balance between lit and unlit spaces to create a visually pleasant and consistent experience. As part of this goal, warm-coloured, glare-free, low-intensity lighting was utilised to meet dark-sky-friendly measures and create a more inviting atmosphere, while the imperative to maximise high energy efficiency ensured that longer-term costs would not be exceeded (Figure 11.3).

The majority of the recently installed lighting at Plas y Brenin is situated below eye level and at door height, locating illumination at a human scale by focusing on the diffusion of the light across the horizontal plane. This decision has also meant

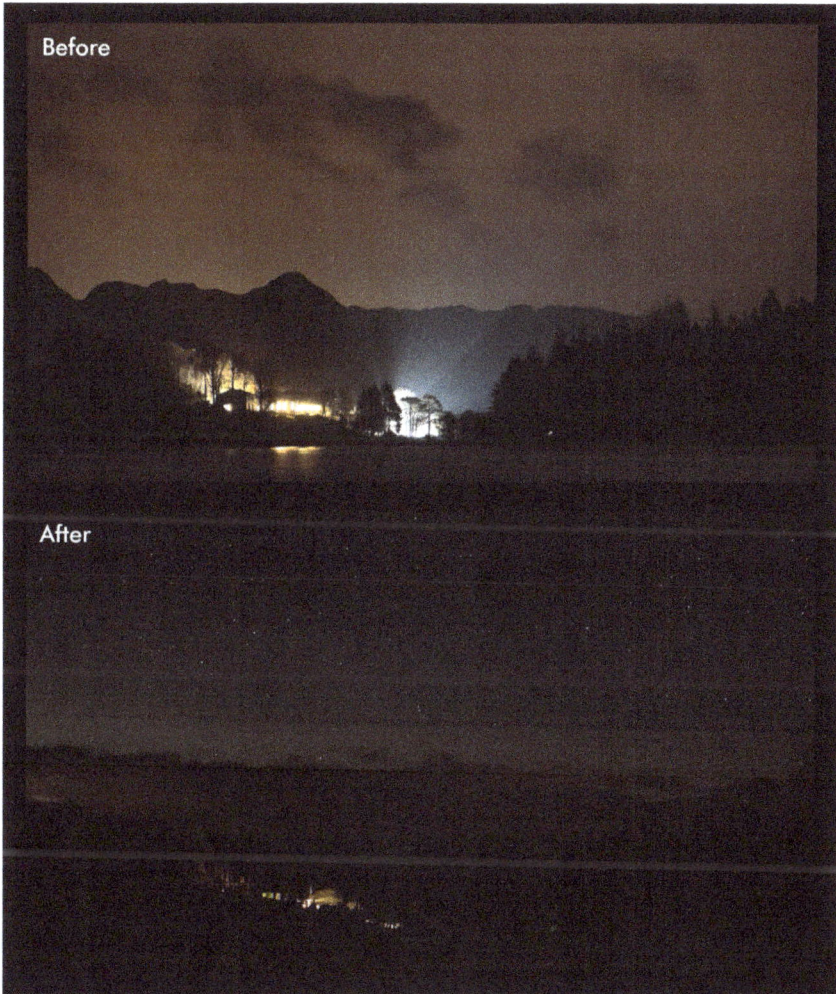

Figure 11.3 Images illustrating the significantly reduced impact of replacement lighting in Plas y Brenin, Snowdonia, Wales. Photos by Dylan Parry Evans, Karl Midlane, and Dani Robertson.

that all lighting is accessible for ease of maintenance and is able to meet high standards of exterior rating and durability. There are also areas in which vertical illumination was essential for reinforcing the legibility of the site in ways that facilitated effective wayfinding, and this was undertaken with the imperative to ensure that luminaires faced downwards to avoid upwards travelling light. Moreover, to create a textured visual character unique to the site, luminaires with elliptical and forward throw were used to increase reach and coverage. Besides the installation of a range of new luminaires, all floodlights were removed from the site. Responsive illumination was also fitted at all the entrance thresholds that are marginal to the flows and circulations of human bodies, with passive infrared sensors (PIR) introduced to avoid unnecessary illumination of these non-essential areas, with lights only coming on after dark and when movement is detected. Where no activity took place, light was not wasted. This use of PIR technology was also extended to certain areas where supplying mains power proved to be challenging, with solar lanterns deployed as a low-energy solution. To fulfil objectives to minimise material and economic waste, existing bollard bodywork was refurbished and maintained so that only light sources were replaced.

The imposition of standardised forms of lighting has often deterred inventive solutions, with specific regulations determining lumen, height, and distribution restricting the capacity of creative, place-specific measures to enhance nocturnal environments. However, at Plas y Brenin, the potential to introduce innovative illumination was facilitated by the fact that the site is privately owned, allowing lighting standards to be partially relaxed. The focus thus moved away from conceiving luminaires as stand-alone items that functionally replace outworn and inappropriate fixtures towards the visualisation of a cumulative effect that promoted the adoption of a holistic perspective that guided how the site would look and feel. This degree of freedom in design also inspired the careful arrangement of multiple layers of light that met dark-sky-friendly criteria and created an inviting night-time experience. With regard to the latter attribute, visitors, who are already attuned to explore darkness through adventure sports activities, are lured outside, advancing the goal of achieving a greater social sustainability by offering people external spaces to inhabit.

Gratifyingly, the new lighting scheme did not only impact upon social, aesthetic, and environmental qualities germane to humans, it has also encouraged the return of bats to the site to forage around hedgerows that were formerly too brightly illuminated. In looking at the larger picture, the project has ensured that a core area of the dark sky reserve is further enhanced and protected, offering a night-time experience aligned with Plas y Brenin's brand and environmental ethos. The overall transition to LEDs has maximised the longevity and energy savings of the scheme through the installation of a wide range of highly efficient lighting products, with energy usage reduced by 1.9 tonnes of CO_2e (8,000 kW) annually. The multi-award-winning project has also served as an exemplar for other environmentally and dark-sky-oriented lighting schemes, and this has been advanced by a recent report on BBC One's *Countryfile* programme that has extended awareness about the important role of considerate lighting design within the dark skies movement.

Newport, Mayo, Ireland

The large, scantily populated areas of Ireland offer great potential for the assignation of dark sky projects. These possibilities have been adopted at Newport, a town with ambitious aspirations to become the first dark sky community in Ireland. Located within the Wild Nephin National Park, which becomes Mayo Dark Sky Park after nightfall, the town aims to establish an important precedent in veering away from rural preconceptions by confirming that populated settlements can also become valuable dark sky destinations. In this regard, Newport Dark Sky Masterplan seeks to inspire other communities to follow suit, by broadcasting its past achievements and future targets in showcasing the full extent of how much can be accomplished through the deployment of considerate illumination.

By the time I became involved, the community has already established a clear vision in aiming to develop a dark sky masterplan with a holistic approach to lighting design that was dark-sky-friendly. The scheme also sought to create an aesthetically enhanced illuminated night-time environment, thus ensuring social and visual sustainability as well as environmental value. As part of this project, two iconic architectural landmarks, St Patrick's Church and the Seven Arches Bridge, were identified as sites at which lighting redesign could eliminate the light pollution and visual discomfort currently caused by their poor illumination.

The first phase of the project focused upon replacing the glaring floodlighting scheme of St Patrick's Church, which was illuminated by luminaires that diminished the beautiful heritage architecture of the 100-year-old building. Visible from numerous points in the town, its hilltop location exacerbated the adverse effects of the poor lighting, with large quantities of wasted glare ascending into the sky without bringing out the ornamental features of the church's facade. Once illuminated by sodium lighting, the floodlighting has been replaced with LEDs, but as in so many cases, this did not properly consider growing awareness that such lighting produces environmental harm to non-humans and humans. The project sought not only to change the lights one by one but reconsider how this project was conceived in order to avoid perpetuating environmental problems in the future.

The project has also been grasped as an opportunity to change the relationship between the church and its surroundings. Before, vertical façade lighting had shaped the visual emphasis in apprehending the church at night, and this was inverted by putting the emphasis on the horizontal ground plane. We have reinstalled lanterns on the railings of the perimeter which has restored the heritage fabric of the church both by day and at night. These lanterns are designed to provide a warm colour temperature (2,200 K) and subtly illuminate the church periphery without causing any glare. This has encouraged the public to visit and explore the church grounds after dark, overcoming the dazzling floodlights which formerly inhibited movement. On the church itself, we explored the potential of emphasising key architectural elements by replacing the formerly undifferentiated illumination of the facades of the building. This was accompanied by the decision to make more substantive use of the interior lighting of the church, which the removal of the glaring façade illumination revealed more acutely (Figure 11.4). In addition,

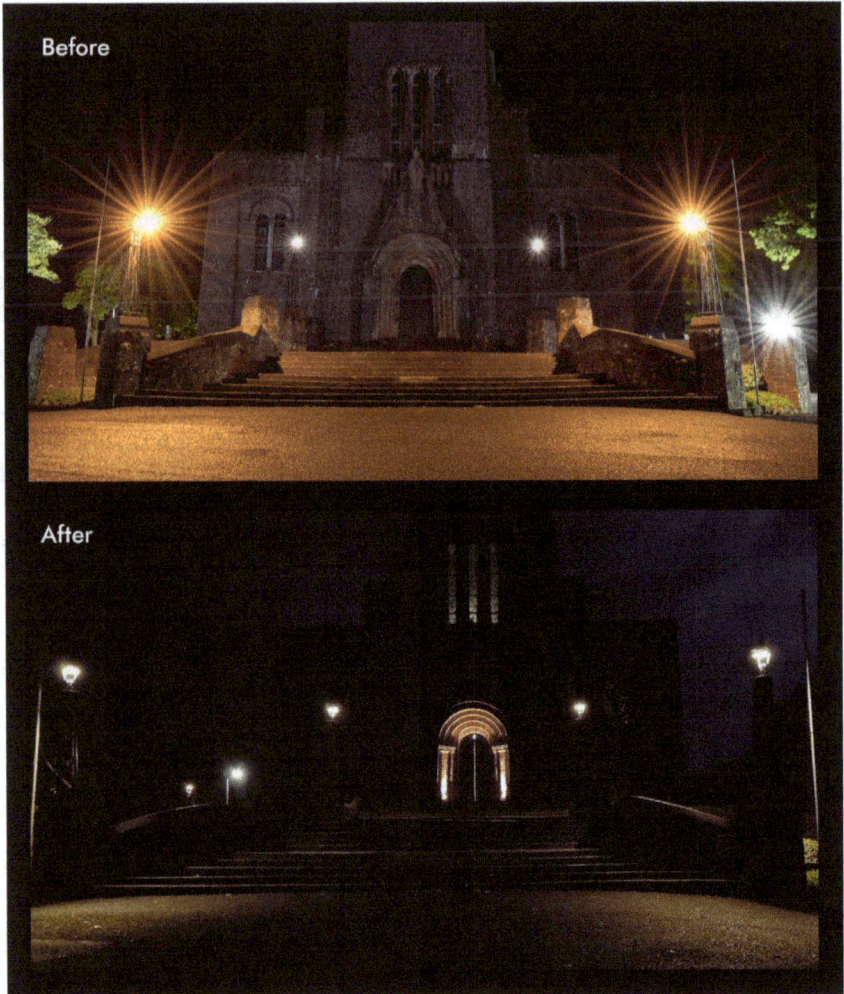

Figure 11.4 Considered lighting for St Patrick's Church makes its different architectural features more distinctive while also providing added emphasis to the ground plane in Newport, Mayo, Ireland. Photos by Georgia MacMillan and Kerem Asfuroglu.

we introduced a switch-off time so that the scheme can allow a period of rest for both the environment and the architecture, and so that the impressive, silhouetted form of the darkened church can stand out against the night sky, adding a different dimension to its visual impact.

Beyond the church, we are currently negotiating the street lighting standards, helping the project to save a significant amount of money and energy while ensuring that the illumination installed is more respectful of the darkness of the night. The imperative to persuade key agencies of the plans, primarily the local authority and the highways authority, has involved communicating the measurements of the existing lux levels which reveal that the current street lighting is excessive for what is required. Following this finding, we have argued for the deployment of a new dimming strategy which would be able to meet a more realistic safety standard as well as offering greater energy saving and reduction in light waste. The highly supportive community of Newport has been critical in this regard by pushing the agenda forward in a dynamic but also diplomatic fashion. This emphasises that by exercising patience and mobilising astute social lobbying, it is possible to influence key powerholders in initiating change. It has thus proved vital to earn the local community's confidence and support in order to reach the decision-makers in the early stages of developing the scheme.

The first phase of Newport's redesigned lighting plan has been delivered. From successfully applying for grant streams to organising dark sky festivals, and in carrying out lighting tests on-site, the community of Newport has a tremendous capacity and willpower to initiate change (see chapter by McMillan *et al.*, this volume). Their participation in devising and installing the lighting scheme—from installing shields on lanterns to learning about how to launch a tender—means that Newport is undoubtedly one of the most strategically minded communities I have worked with. As a consequence, not only has the project improved the night-time experience of the town, but we have also presented evidence to the county council and residents with the aim of securing further dimming measures across the town, a measure estimated to save ten tonnes of CO_2 per year.

Conclusion

Since illumination invariably has environmental, social, and economic costs, certain priorities in light design for dark skies focus on particular qualities, notably the use of warm colour temperatures, mechanisms to shield the light from spreading vertically and horizontally across space, the elimination of glare and excessive brightness through the deployment of lower intensity luminaires, and responsive technologies that dim and switch off lights when appropriate. Perhaps the most significant question is whether light is needed at all in particular settings and locations (ILP, 2021). The ability to install advantageous deployments, however, depends upon the light designer's degree of autonomy and influence, especially in working with local authority approval with regard to sidestepping the specific lighting standards that regulate the qualities of light that shine on streets, pathways, and squares. Consequently, it is critical to strategically target such authorities and gain their confidence rather than confronting these gatekeepers prematurely; while this may be time-consuming, influence and goodwill are apt to eventuate in the fostering of joint initiatives.

In this context, it is timely to remember that such official strictures are not necessarily embraced by all and that they are always susceptible to change inspired by new ways of thinking, feeling, and making. For instance, in 17th-century early modern Europe, when municipal lighting was first introduced, it was widely understood as an authoritative strategy that intended to police the night rather than encourage social activity. Rebel acts of lantern-smashing were celebrated as resistance to these desires for control (Schivelbusch, 1988). Perhaps the lofty, somewhat inaccessible streetlights of today inherit some of their characteristics from the days when height was used both to deter vandalism and illuminate an area as widely as possible. Battles over the ownership of the night, symbolised by these acts, persist, and centuries later, communities are still fighting against the ubiquitous nature of municipal lighting (Commission for Dark Skies, 2020). More recently, one of the key ways in which dissatisfaction with the over-illumination that has culminated from a regulatory approach has been articulated through expanding demands to express the right to darkness.

The three examples provided in this chapter highlight how designing illumination schemes for dark sky places must be guided by specific, place-based aesthetic and material qualities, environmental imperatives, social agents, and political contexts. I have demonstrated that a host of technologies and attributes need to be considered in deciding what kinds of luminaires to install and choosing where light should and should not shine, as well as informing the choice of qualities of radiance, colour, and intensity. The development of improved LED technologies and responsive lighting has assisted in the devising of effective environmental and aesthetic objectives. As manufacturing costs are reduced, new modes of marketing emerge, more sustainable technologies are developed, and the range of aesthetic potentialities expand, possibilities for creating diverse lighting schemes are expanding. Yet much inequality endures in the distribution of light across the world. Some underprivileged communities are bombarded with surveillant light while others struggle to obtain the most basic illumination. In the UK, while this inequality persists, movements to develop more sustainable, aesthetic, convivial, and place-specific lighting are gaining traction.

The ongoing testing of variations in luminosity and placement of luminaires that collectively composed columns of light in Presteigne fully involved local residents in contributing to the evolution of the proposed scheme. The privately managed site of Plas y Brenin allowed a greater degree of freedom in design, sidestepping some of the regulatory impositions that curtail lighting schemes in the public realm. This project allowed for a greater range of luminaires to be deployed as part of a holistic approach to honouring the dark skies of the Snowdonia National Park. And the hilltop location of Newport's St Patrick's Church means that its formerly over-illuminated façade contributed to an excessive skyglow and the radiation of unnecessary quantities of illumination across space that could be seen from numerous angles. This particular intrusive, site-specific problem needed to be dealt with as a signal to the community and to the authorities to dramatise the effects of installing more sustainable, aesthetically appropriate lighting.

In all the three cases discussed here, highly supportive communities who have been patient and committed to achieving project goals have been essential in achieving successful outcomes, as have local authorities who have proved to be flexible and supportive in most cases. The cultivation of relationships and connections to the communities and to key players has been critical in encouraging their understanding. Indeed, the engagement with the local community has been the most critical factor in developing the schemes discussed here. It has proved essential for these communities to be attuned to a dark skies ethos, a perspective that may predate the engagement of the designer or which can be developed through sustained discussion and the sharing of objectives.

Dark sky destinations act as cultural beacons in broadcasting the values of the movement to a wider audience. Yet while restoring dark skies initiates a shared project, keeping such realms dark is a further challenge. In this context, it is crucial to provide agency, ownership, and voice to people to participants in order to nurture a duty of care and responsibility. Light pollution is relatively easy to rectify, and speedy results empower communities who come together to create forms of illumination that allow darkness to prosper. Yet for some, notably those who are located close to brightly illuminated urban areas, dark sky environments are far less easy to achieve. This discloses that it is vital that the dark sky movement spreads to urban communities as well as the more rural or small-town settings discussed here.

For the ultimate frontier of the battle for dark skies are cities. Much work is still required in order to demonstrate that urban areas can also be installed with environmentally friendly lighting. For instance, a perusal of a map of London discloses that its largest outdoor assets are its parks, spaces that are ordinarily unilluminated after nightfall. The designation of urban night sky places as a potential tier created by dark skies has been designed specifically to encourage urban communities to take action by firstly protecting the darkness that pervades these recreational spaces before seeking to expand the area in which darker environments might be designed. Illumination has not been part of an evil plot to colonise the night; rather, layers of light have accumulated over time without being paid attention and because of a lack of awareness about the malign effects that can eventuate. Now we know of the aesthetic, biological, environmental, and social harms that are produced by excessive and poor illumination, there seems little excuse to ignore the situation. The opportunities to devise lighting schemes that are informed by the benefits of dark skies provide a useful way to develop more varied, more sustainable, and more aesthetically pleasing nocturnal environments, as the examples provided in this chapter demonstrate.

References

The City Dark. (2011) [Film] Dir. Ian Cheney, USA: Wicked Delicate Films.

Commission for Dark Skies. (2020) *Blinded by the light? A handbook on light pollution.* Wimborne: British Astronomical Association and Commission for Dark Skies.

Commission for Dark Skies. (2021) *Hidden stars: Where have the stars gone?* Wimborne: British Astronomical Association and Commission for Dark Skies.

Crary, J. (2013) *24/7: Terminal capitalism and the ends of sleep.* London: Verso.

Dark Source & Friends of the Lake District. (2023) *Designing out light pollution in Cumbria and Arnside & Silverdale AONB: Good lighting technical advisory note.* London: Dark Source.

Ekirch, R. (2005) *At day's close: Night in times past.* London: W.W. Norton and Company.

ILP. (2021) *Guidance note 01/21 the reduction of obtrusive light*. Rugby: Institution of Lighting Professionals.

Koslofsky, C. (2011) *Evening's empire: A history of night in early modern Europe*. Cambridge: Cambridge University Press.

Schivelbusch, W. (1988) *Disenchanted night: The industrialisation of light in the nineteenth century*. Oxford: Berg.

12 Who is afraid under dark skies?

Four female experts on "spaces of fear", astronomy, and the loss of the night: a group discussion with Sabine Frank, Josefine Liebisch, Laura-Solmaz Litschel, and Dunja Storp

Nona Schulte-Römer

Introduction

Recently, at a friend's house, I found myself reproducing female fears of darkness. When I met their guest, a young lady from Mexico who was just about to go for a run in the dark, I could not help asking, "Are you not afraid?" She seemed surprised by my question, joked about how safe Berlin was compared to Mexico City—and off she went into the dark.

I am not the only one asking such presumptuous questions. There is an influential discourse that problematises the darkness as a safety risk, even though it is so essential for life on earth. Nocturnal darkness allows our bodies to find rest if we let them and orchestrates the natural rhythm of ecosystems (Gaston, Visser and Hölker, 2015). Starry skies have always fascinated humans and allowed premodern civilisations to make their way around the night as farmers, sailors, and storytellers. The stars have been a key to human progress and culture. Yet today, the experience of dark night skies has become a rarely experienced good and nocturnal human activities have diversified. In cities, the night is illuminated for urban night-time economies and night work keeps logistics and mobility infrastructures running and saves people's lives (Dušková and Duijzings, 2022). Human night-time activities have to date coincided with a relentless and increasing hunger for artificial lighting. As a result, the Anthropocene is also the epoch in which we are "losing the night", as astronomers, environmentalists, citizens, and even lighting professionals are demonstrating (Rodrigo-Comino *et al.*, 2021; Schulte-Römer *et al.*, 2019). A key concern is that too bright and too numerous lighting fixtures are being deployed in the name of road safety and personal safety, especially the safety of women.

This chapter tackles such safety-first arguments from a gendered perspective. In the following, I present a conversation between four female experts of dark nights and skies. While their professional and disciplinary backgrounds vary, all four are from Germany, so they share some sociocultural contexts. The edited short version of our discussion has been translated into English and touches upon "sham debates" about darkness, light, and urban angst, as well as "female astronomy".

DOI: 10.4324/9781003408444-17
This chapter has been made available under a CC-BY-NC-ND 4.0 license

Figure 12.1 The four interviewees, from left to right—Laura-Solmaz Litschel, Dunja Storp, Josefine Liebisch, and Sabine Frank. © L.-S. Litschel/M. Bustamante; D. Storp; Liebisch/T. Mittmann, and S. Frank.

Experts of the dark: from nocturnal research to guided stargazing

Why do dark skies matter? My four participants in conversation answer this question from various perspectives. While the amateur astronomers Josefine Liebisch and Sabine Frank look up into the dark sky to observe the stars, Laura-Solmaz Litschel and Dunja Storp study night work and nocturnal behaviour on the urban ground. Here, I provide a closer look at their night-time activities and how they relate to dark skies. Wherever they referred to studies or literature, I have added these references as endnotes.

Dunja Storp is a criminologist and urban safety consultant with an interdisciplinary training in architecture and the social sciences. She also knows police work from the inside, having worked for the police in her previous career. Her clients, mostly municipalities and planners, consult her to prevent crime and improve urban spaces that are deemed unsafe. Her work includes the study of all relevant data—crime statistics, socio-demographic data, structural and architectural facts—and site visits at different times of the day, which allow her to understand why particular urban spaces are considered problematic.

Dunja: In my criminological spatial analyses, I look at very small areas. I prefer to work in areas that concern artificial lighting. I look into night-time spaces in which people develop fears, where people seem to behave in anti-social ways. I go there at night and ask about what could be the reason for this kind of behaviour and how can we influence it? I have to know exactly what happens and where in a space to put it into context. This small-scale approach is important as criminological research shows that almost 80 to 90% of crimes committed in cities can be limited to 5% of their urban spaces (Braga and Weisburd, 2010). There are reasons for why drug dealing happens in a certain place and not in another. But crime statistics do not offer that information. I need to get a personal impression and *feel* the space to understand it. Based on my analysis, I make suggestions on how such places could be designed and illuminated differently.

For our joint discussion at 10:00 a.m., Dunja got up earlier than usual after spending another night outside in an uncanny urban space.

Laura-Solmaz Litschel studies night-time activities in urban spaces as part of her social-scientific research on digital labour in the transnational project "Night spaces: migration, culture and integration in Europe" (NITE), at the Humboldt University of Berlin. During her research, she interviews people who deliver food or take care of e-scooters throughout the night, most of them male. She also observes and talks to people in the night-time streets of Berlin and lingered at e-scooter stations. In her doctoral thesis, Laura explores how digitally organised platform work in Berlin transforms not only work relations and rhythms but also the urban night. She thereby engages critically with smart city imaginaries and the "myth" of the "24/7 city" (Henckel, 2019). To offer some context, in Germany 4.2% of the working population regularly work at night (between 11:00 p.m. and 6:00 a.m.), which is also the average percentage for the 27 EU countries. Men work nights almost twice as often (5.4%) as women (2.8%) (Eurostat, 2023).

Laura: We study 24/7 urban internet services and when we look more closely at night-time working realities, we see that the urban night still functions differently than the city by day. The night worker does not necessarily meet other people in the streets, but he meets a fox! Nocturnal animals are part of the urban night as a social phenomenon. Then, of course, people need other tools to work at night, like their smartphone torches and streetlights. As artificial light is expanding, this kind of digitally supported nightwork is spreading. The two are mutually dependent.

An especially striking night-time impression from Laura's night-time fieldwork is "the glow of an e-scooter in the dark water of the Spree River". An interview partner had to fish the scooter out of the river after vandals had dumped it there.

Josefine Liebisch published her first astronomy paper on variable stars—that is, stars that change their brightness—while still in high school. Today, she studies arboriculture in Göttingen, Germany, and is part of the volunteer team that runs the Gönnsdorf observatory in Dresden.

Josefine: We work a lot with schools in the region. We also supervise student projects. At the moment, we have a student who is observing an exoplanet transit. That is, when another planet outside our solar system passes in front of its star. What you see from the transit is that the planet moves in front of the star and in this case, obscures it by 2%. And we can measure this dimming. That's really crazy because the discovery and observation of these extrasolar planets is really something for space missions or for proper scientific institutions rather than amateur observatories.

Once a year, the Gönnsdorf observatory team travels to stargazing weekends that take place in various places across Germany. Participants in these meetings come with binoculars and only observe with their naked eyes. Others "arrive with their high-tech telescopes and cameras in small vans", explains Josefine.

Sabine Frank is an amateur astronomer too but also Germany's first official dark night advocate (Nachtschutzbeauftragte), working for the UNESCO Biosphere

Reserve Rhön and the County of Fulda. In 2009, Sabine started campaigning for the Rhön to become a dark sky reserve, a goal that was achieved in 2014 (Frank, 2023). "The issue of light pollution has been my concern ever since", she says. At daytime, she coordinates the nocturnal activities in the reserve and offers advice on dark sky protection to both public and private actors. She has attended "an incredible number" of local council meetings, contributes to local development plans and national bylaws and has developed a label for "light awareness" to incentivise private companies to illuminate less. At night, she takes young and old visitors on nocturnal walking tours to show them the beauty of natural darkness and starry skies. In the past two years, she estimates that she has guided about 150 tours.

Sabine: Although I live in the countryside, where you can actually see the stars quite well, people think they need technical equipment to look into the sky. In 2006, I started simply walking through the night with people. Amateur astronomy is often about observing deep sky objects, exposing photographs for many hours and post-processing them to show them in social media. My perception of the night is different. It includes the night landscape as well as the stars. With my star guiding tours, I wanted to de-technicalise stargazing. I call it "female astronomy".

While Sabine is tirelessly lobbying for dark skies, she is very tired of light pollution and opposes the claim that women prefer illumination over darkness.

Gendered engagements with dark skies on trial

When it comes to protecting dark skies, key counterarguments are articulated around gendered assumptions about light and safety prioritising female demands for lighting. We discussed these contentions in depth and partly through opposing perspectives. While Dunja and Sabine see a need for more female perspectives in urban planning and amateur astronomy, Laura argues against binary gender categories, and Josefine enters our discussion with the very legitimate question of why we discuss these issues amongst women only: "I think it makes sense to talk to men as well, so that arguments come from different perspectives and not just from the women's point of view". In this sense, we invite all readers to engage with our discussion and give us feedback. What we propose on the following pages is a more nuanced understanding of not just female but human fears. Furthermore, we propose to reconsider urban planning for night-time cities, the relationship of light and safety, and our engagements with the dark, starry sky.

Female fears in the dark?

Although fearful feelings in the dark are often ascribed to and reported by women, it is important to acknowledge that darkness enhances different sources of insecurity. Visually oriented as we humans are, we feel vulnerable if we cannot see our environment. Social fears make things worse since we also cannot see who is lurking in the dark. For those who would rather run than fight, the loss of visual orientation can stir up fear of victimisation worse. Luckily, there are remedies: the social fear of crime is alleviated in the presence of trustworthy fellow-people, and space-related fears of disorientation subside in familiar environments. Artificial light offers a particular powerful solution as it allows us to spot both potential offenders and helpers and to better navigate through unfamiliar surroundings. Yet, gendered assumptions often miss the universality and complexity of these fearful feelings. For although women are considered most vulnerable in public spaces at night, men fall victim more often, emphasises Sabine (Birkel *et al.*, 2022). Moreover, light is not always better. Women also report that they actually feel safer under the veil of darkness where they are not on display and invisible to potential offenders. Indeed, the night has always been a protective space for socially marginalised groups (Ekirch, 2005).

Laura: Certain social groups experience the night as a space of protection. If we look at who works in cities at night, we often see people who work in precarious jobs and belong to socially marginalised groups, such as people with a history of migration. Also, the night has always been a protective space for trans and queer people and for sex workers.

Still, darkness can also stipulate fearful thoughts, as the lack of visual cues allows our imagination to run wild and to imagine potential dangers in the dark, as Josefine recalls.

Josefine: Although I go outside at night to look at stars, I'm sometimes afraid of the dark. I've maybe watched too many horror movies. For instance, at our observatory we have an outdoor space that lies next to a field. One night in late summer, we were stargazing there. The corn on the field stood high and suddenly it rustled directly at the fence. I completely panicked—even though I knew, it was completely senseless to be afraid, that it was just a wild boar. In the end, it was funny.

Sabine thinks that fears of darkness are fostered by all sorts of popular stories and media images and that such narratives should therefore be reframed:

Sabine: I also had a scary experience once. That's why you have to cast darkness positively, as a natural and calm space, as something inspiring and fascinating. In crime novels, women are murdered in the woods after dark. In reality, people are not murdered in the forest at night! This distracts us from where the violence against women actually takes place. It disgusts me that every third day a woman is murdered by her partner behind her own apartment door!

In public debates however, it is mostly uncanny urban spaces after dark that (re) produce social fears of crime, particularly female fears.

Urban angst

In German urbanism, the notion of Angsträume ("spaces of fear") and the term Dunkelräume ("spaces of dark") are exemplary for political and planning debates. As a night-and-safety expert, Dunja looks at "everything that bothers citizens" and has come to the conclusion that it does not need darkness and actual crime to upset people. Trash on the street, graffiti on walls, and especially uncivil behaviour suffice to turn public spaces into uncanny Angsträume (Doran and Lees, 2005). Women then tend to avoid these places.

Dunja: People are very, very different and there are many influencing factors. But it can't be refuted—and studies show this over and over again—that women are more likely to talk about fears in public spaces. They often tell me, "I don't go in there, there's a lack of light". If I then ask specifically where light is missing, they have no first-hand experience but tell me what they know from the media about "no-go areas" or from friends. Men, in my experience, tend to be more reluctant to speak out about their fears.

Dunja criticises how "urban spaces are almost exclusively shaped by male planning perspectives" with a focus on cars and technological solutions. With a "female perspective", which she finds wanting, Dunja associates urban planning and design that reflects the actual needs of families and elderly people rather than justifying designs, technology choices, and demands for "always a lot of light" with reference to women's fears. Dunja further criticises the process through which women "learn from an early age to avoid certain spaces like train stations or parks because you just don't go there. At night-time, you shouldn't be outside alone as a woman". It seems that urban spaces at night are not for everyone.

Dunja: When I'm out and about in the cities at night, I hardly meet women alone or any elderly people. Maybe on the weekends. The women I see tend to scurry through the public spaces. I find that frightening. But I do see many young men who have no other social spaces to hang out.

Yet the absence or shyness of women in night-time public spaces cannot be transferred one-to-one to other urban spaces or even rural areas, as Laura notes.

Laura: When we travel through Berlin at four in the morning on public transport, we see many people who I read as women. Many of them work in the cleaning sector, some through digital platforms and apps. That we can access goods and services 24/7 through the internet does something to the night and specifically to urban nightlife. In the countryside, you can't access these services to the same extent.

This matches with Sabine's experiences in the Fulda area where the population density is 25 times lower than in Berlin (Statista, 2023; Wikipedia, 2023). "We don't only have Berlin and the Hamburg Reeperbahn", she explains. Cities differ and in places like Fulda, the streets are nearly empty at 9:00 p.m. In the countryside, Sabine meets even fewer pedestrians. "You don't see any people walking around—and it doesn't matter whether it is 7:00 p.m. or 9:00 p.m. They only drive in their cars and these cars have headlamps. Nonetheless, public lighting shines and shines and shines and shines for nothing and nobody".

"Sham debates" about light and safety

In the last decade, light and safety debates seem to have reached a new level. Since energy-efficient light-emitting diodes (LED) have entered the market, lighting consumes less energy and has become cheaper. This has led to rebound effects and an increase in light emissions.

Sabine: Before the LEDs, when light was still perceived as an expensive resource, it was immediately clear to everyone that you don't use it wastefully. I can remember a mayor who considered it common sense to put as little light as possible on his streets. Crime was not an issue at all. At most, politicians discussed traffic safety. Since then, the situation has changed dramatically. Now that LED lighting is cheap and has also been subsidised by the EU, over-lighting is increasing.

The criminologist, too, is critical of lighting as a solution to security issues. Dunja tells how she just visited an urban space which her client had described with the words: "We need more light. They're smearing all over our facades with their graffiti". Yet when she arrived, she found the place and especially the graffiti perfectly well illuminated and wondered: "Why don't you turn off the lights, then you won't have this graffiti problem".

Dunja: In areas with social unrest, security lighting has been introduced. It now prevents people from entering these spaces because they feel blinded. It makes people feel illuminated and insecure. Then there is the justification that "the police need light". That is as right as it is wrong. The police need light when they record an accident and secure evidence. But if you talk to police about rioting groups, they say, "When the lights are turned up, it's like a switch is flipped. Then the trouble really starts".

This activating potential of artificial light at night is well researched, explains Dunja, and is caused by physiological hormonal processes that can differ depending on people's gender, age, and socio-cultural perceptions. "One example is that women find cold white light sources much more unpleasant and disturbing than men". Other studies show that men respond more aggressive to bright light. Moreover, lighting in the early evening has different effects than late at night. Dunja concludes that "we are not wired to be active at night" and criticises situational crime prevention that does not sufficiently account for these biological responses. Although Laura and I have doubts about biological arguments that can also reinforce gender differences in political debates, we all agreed that lighting solutions are no panacea for societal conflicts and that public debates are often misplaced.

Sabine: These debates about "spaces of fear" and "dark spaces" are completely urban and not suited for the countryside, where it is commonly dark at night. The night seems a great topic for polarising and for election posters with slogans like "More light means more security". As if there are no real problems. We have massive species extinction in open natural spaces outside of protected areas. We need urban ecosystems—also at night, because they are refuges for quite a lot of animals. It makes me sad and I'm at my wits' end because I don't know how to break through these sham debates. We have a lighting industry that lobbies for more and more light and government ministries that go along with it. They disrupt the night, even though we all find it inspiring and calming. I meet so many people who think starry skies are awesome and also women who would like to have less light.

Instead of calling for more lighting, we should ask, "What is good lighting and at what time?" concludes Dunja. "Darkness is perfect at times".

"Female astronomy"? From male high-tech astronomy to dark skies for all

Increasing light emissions both in terms of light levels and expanding illuminated areas have already considerably reduced people's chances to experience dark starry skies (Kyba *et al.*, 2023). In light of the long legacy of sky observations and astronomy, this is also a cultural loss. We thus also explore the value of dark skies and stargazing practices which, despite their universal appeal to humans, are not free from gender issues either. In Josefine's observatory, only four out of 21 volunteers are women, including the head of the association, who "keeps the place together", she reports. Sabine sees a clear connection between technologically enhanced astronomy and gender.

Sabine: Fifteen years ago, amateur astronomy was mostly male—if only because the telescopes used to be really heavy and difficult to adjust. I think the focus on technology and always utilising bigger and bigger telescopes has alienated people from simply looking at the sky with the naked eye. Many believe you really need a telescope to see Jupiter, but stargazing should not be left to telescopes and photography. The night and the starry sky belong together, and their observation should not be limited to a few observatories and remote places. I know astro clubs that fly to Namibia once a year. They do that regularly! As a result, astronomy is expensive, often white and male. At the same time, they don't campaign against light pollution here at all.

Sabine's star tours are anything but "technicised". Although she carries binoculars with her, the focus is on simply walking in the dark for two hours so that the participants can appreciate "how a tree stands out against the background of the sky" and tune in with the darkness of nocturnal landscapes. Apart from enjoying starry skies, the educational goal is that participants experience the deployment of their own visual abilities at night and learn more about dark sky protection.

Sabine: We walk everywhere in the dark, even on stony paths. No one has ever fallen. We can see exceptionally well at night—but have lost confidence in this ability. I'm surprised how often people tell me, especially people from the city, that they didn't know that they could walk outside without lights. How alienated have we become from relying on natural night light!

Sabine's proposition of a "female astronomy" does not remain unchallenged in our discussion. Laura finds it "important not to equate men with technology and women with a somehow natural access to the night". While experiences of the night might indeed differ across social groups, she prefers to consider these differences as an outcome of social structures. Josefine is impressed by the concept of star tours, but also questions the idea of a male, technicised astronomy. In their observatory they engage with dark skies and the stars in multiple ways—with the naked eye, with binoculars, or with a telescope. There are only two colleagues who "take ten hours of photos and then spend another 30 hours on photo editing", Josefine tells us, laughing.

Josefine: We reach all age groups, men and women, and especially children and young people. There are kindergarten children who ask you about black holes and why the universe is infinite and yet has limits. Then you stand there and ask yourself: kid, how do you know that? Of course, these questions are not easy to answer.

In her "rebellious phase", however, she has developed a lecture about "women in astronomy".

Josefine: Women in astronomy are simply not well-known. When it comes to great discoveries, most people think of men because discoveries are named after men. So I asked, what did women do? Lo and behold, they existed and did just as much important work as men! Over the years, my perspective has shifted: Why should I exclude men from the story? I now say in my lecture, let's look at a scientific discovery and see who worked on it—women as well as men.

Finally, all four experts share a deep appreciation of dark skies and agree that the proceeding "colonisation of the night" should be stopped (Melbin, 1978; Koslofsky, 2011).

Sabine: The destruction of darkness is life-threatening. We get along just fine with moonlight. I want people to see darkness as a quality of life when they sit on their balcony at night. I have succeeded in getting many communities to turn off their streetlights and they no longer feel that something is being taken away from them.

Dunja: We need the night. We are living beings with a natural rhythm. Working against it doesn't do us any good. When I end my work at night, I enjoy just sitting down in the dark, somewhere outside and looking up to come to myself, to calm down and get tired. I need the night sky to ground me before I go home.

Josefine: My experience is that stargazing brings people back down to earth. It makes us realise that we are just a tiny part in a much bigger picture. In this larger context, I know that I am just one person amongst many and I live on one planet of many. I wish that everyone could have this experience, simply because it helps us to broaden our limited, understandably human view and to open up to what is there above us.

Laura: Everyone needs darkness. But through new forms of work, lighting, mechanisation, of course, we have ways of going further and further into the night and turning it into day. That relates to a commons debate that's coming up now that we're also already aware of.

This debate about "nocturnal commons" (Crary, 2013, p. 5; Gandy, 2017; Dunn, 2020) offers a pertinent outlook to our discussion as Laura highlighted a newly emerging risk: As Elon Musk is shooting Starlink satellites into the sky, they add to light pollution as they are reflecting the light back and amplifying skyglow (Shepherd, 2023). Laura and Josefine therefore see a need not only for transnational governance of artificial light on earth, but also for regulating human activities in outer space.

Conclusion: dark skies as "cosmic commons" and a cosmopolitical proposal

In light of our discussion, there seems an urgent need for widening our perspective and not only discuss "nocturnal" but "cosmic commons"—in resonating with philosopher Isabelle Stengers' (2015, pp. 994–995) "cosmopolitical proposal". This proposal promotes a notion of politics that allows a "cosmos", a "good common world", and invites us to slow down and create spaces for hesitation and discussion about what it means to say "good". In this sense, the notion of cosmic commons both includes and goes beyond nocturnal commons. It draws attention to recent knowledge about the value of darkness for living species, including humans, ecosystems, the various ways in which individuals and cultures make sense of urban night and darkness, the differences between countryside and city nights (Edensor, 2017). Yet it also implies tensions. Existing lighting demands and fears of darkness that are shaped by cultural practices and histories (Le Gallic and Pritchard, 2019) inevitably clash with the more recent call to protect dark skies. In this sense, a "cosmopolitics of dark skies" invites trouble and negotiations as it would resist the temptation of illuminating the night as a standard operation or sign of modernity and progress. Instead, it challenges us to rethink our lighting practices in light of conflicting values, new knowledge about darkness, and new technological options. It also means that we might find a stance to slow down the colonisation of the night and deconstruct misleading generalisations.

In our two-hour discussion, we made a start by challenging taken-for-granted assumptions and simplified arguments about darkness, light, and safety with a focus on gender issues. We found that walking confidently in the dark is a matter of habit and can be learned. Female fears are human fears and are enhanced by the lack of vision in the dark, media-shaped imaginations and social dynamics like the presence of strangers in wary-looking places or Angsträume. Illuminating these places makes it easier to detect potential threats but can make things worse—by blinding people, creating extra-dark corners, and exposing vulnerable social groups even more, irrespective of their gender (Üblacker and Lukas, 2023).

Hence, and cosmopolitically speaking, generalised assumptions about female fears are problematic not only because they reproduce gender prejudices but also because they distract attention from fellow creatures—humans, diurnal and nocturnal animals, and plants—that have their own needs for darkness and protection, but maybe no voice to speak for themselves. Moreover, there seems to be an increasing number of women who do speak for themselves and call for darker skies. Obviously, a cosmic commons perspective is at odds with existing standards for traffic safety, crime prevention routines, excessive night-time economies, or work safety in nocturnal airports and logistic centres. Stengers' cosmopolitical proposal is a call to accept and deal with these tensions. Hence, a cosmopolitics of dark skies recognises the multiplicity of perspectives, confusing or even contradictory facts, incommensurable values, and incompatible views in order to struggle for "good" solutions and protect our cosmic commons.

In practice, cosmopolitics might include very simple steps, which brings me back to my initial example of female fears and running in the urban dark. Last autumn, British former world-champion cyclist and governmental spokesperson Chris Boardman (2022) publicly discussed in *The Guardian* why his wife did not run in the dark and ended with "Calling all men: this is what we can do to help women feel safe exercising in the dark". Boardman urges men to not run closely behind women, stop making unsolicited compliments or advances, and spread the word. Not once does he suggest that the lights should be turned up for female runners.

References

Birkel, C., Church, D., Erdmann, A., Hager, A. and Leitgöb-Guzy, N. (2022) *Sicherheit und kriminalität in deutschland—SKiD 2020*. Wiesbaden: Bundesweite Kernbefunde Des Viktimisierungssurvey Des Bundeskriminalamts Und Der Polizeien Der Länder. Available at: www.bka.de/DE/UnsereAufgaben/Forschung/ForschungsprojekteUndErgebnisse/Dun kelfeldforschung/SKiD/Ergebnisse/Ergebnisse_node.html (Accessed: 17 April 2023).

Boardman, C. (2022) 'Calling all men: This is what we can do to help women feel safe exercising in the dark', *The Guardian*, 30 October. Available at: www.theguardian.com/commentisfree/2022/oct/30/women-safe-exercise-in-the-dark (Accessed: 17 April 2023).

Braga, A. A. and Weisburd, D. (2010) *Policing problem places: Crime hot spots and effective prevention*. Oxford: Oxford University Press.

Crary, J. (2013) *24/7: Terminal capitalism and the ends of sleep*. London: Verso.

Doran, B. J. and Lees, B. G. (2005) 'Investigating the spatiotemporal links between disorder, crime, and the fear of crime', *The Professional Geographer*, 7(1), pp. 1–12. https://doi.org/10.1111/j.0033-0124.2005.00454.x.

Dunn, N. (2020) 'Dark design: A new framework for advocacy and creativity for the nocturnal commons', *The International Journal of Design in Society*, 14(4), pp. 19–30.

Dušková, L. and Duijzings, G. (eds.) (2022) *Working at night: The temporal organisation of labour across political and economic regimes*. Berlin: Walter de Gruyter. https://doi.org/10.1515/9783110753592.

Edensor, T. (2017) *From light to dark: Daylight, illumination and gloom*. Minneapolis, MN: Minnesota University Press.

Ekirch, R. (2005) *At day's close: Night in times past*. London: W.W. Norton and Company.

Eurostat. (2023) *Employed persons working at nights*. Available at: https://ec.europa.eu/eurostat/databrowser/view/LFSA_EWPNIG__custom_4841139/bookmark/table?bookmarkId=df90f49a-940b-4778-adab-20b6b2ededa8 (Accessed: 17 April 2023).

Frank, S. (2023) *International dark sky reserve "Rhön"*. Available at: www.biosphaerenres ervat-rhoen.de/natur/sternenpark-rhoen/international-dark-sky-reserve-rhoen (Accessed: 17 April 2023).

Gandy, M. (2017) 'Negative luminescence', *Annals of the American Association of Geographers*, 107(5), pp. 1090–1107.

Gaston, K. J., Visser, M. E. and Hölker, F. (2015) 'The biological impacts of artificial light at night: The research challenge', *Philosophical Transactions of the Royal Society B: Biological Sciences*, 370(1667), p. 20140133. https://doi.org/10.1098/rstb.2014.0133.

Henckel, D. (2019) *The 24/7 city—a myth*. Available at: http://unsichtbare-stadt.de/the-24-7-city-a-myth/ (Accessed: 17 April 2023).

Koslofsky, C. (2011) *Evening's empire: A history of night in early modern Europe*. Cambridge: Cambridge University Press.

Kyba, C. C. M., Altıntaş, Y. Ö., Walker, C. E. and Newhouse, M. (2023) 'Citizen scientists report global rapid reductions in the visibility of stars from 2011 to 2022', *Science*, 379(6629), pp. 265–268. https://doi.org/10.1126/science.abq7781.

Le Gallic, S. and Pritchard, S. B. (2019) 'Light(s) and darkness(Es): Looking back, looking forward', *Journal of Energy History/Revue d'Histoire de l'Énergie*, 2. https://energyhistory.eu/en/node/137.

Melbin, M. (1978) 'Night as frontier', *American Sociological Review*, 43(1), pp. 3–22. https://doi.org/10.2307/2094758.

Rodrigo-Comino, J., Seeling, S., Seeger, M. K. and Ries, J. B. (2021) 'Light pollution: A review of the scientific literature', *The Anthropocene Review*. Online first. https://doi.org/10.1177/20530196211051209.

Schulte-Römer, N., Meier, J., Dannemann, E. and Söding, M. (2019) 'Lighting professionals versus light pollution experts? Investigating views on an emerging environmental concern', *Sustainability*, 11(6), p. 1696. https://doi.org/10.3390/su11061696.

Shepherd, T. (2023) 'Picture imperfect', *The Guardian*, 5 January. Available at: www.the guardian.com/science/2023/jan/06/picture-imperfect-light-pollution-from-satellites-is-becoming-an-existential-threat-to-astronomy (Accessed: 17 April 2023).

Statista. (2023) *Bevölkerungsdichte in Berlin von 1995 Bis 2021*. Available at: https://de.statista.com/statistik/daten/studie/255791/umfrage/bevoelkerungsdichte-in-berlin/ (Accessed: 17 April 2023).

Stengers, I. (2015) 'The cosmopolitical proposal', in Latour, B. and Weibel, P. (eds.) *Making things public: Atmospheres of democracy*. Cambridge, MA: The MIT Press, pp. 994–1003.

Üblacker, J. and Lukas, T. (2023) 'Local cultures of control, order maintenance policing, and gentrification: A comparison of railway station districts in three German cities', *Journal of Urban Affairs*, pp. 1–24. https://doi.org/10.1080/07352166.2023.2174870.

Wikipedia. (2023) *Landkreis fulda*. Available at: https://de.wikipedia.org/wiki/Landkreis_Fulda. (Accessed: 17 April 2023).

13 What do we mean by "dark skies"?

Yee-Man Lam

Introduction

My friend travelled to the Sahara wishing to enjoy a star-filled sky; unfortunately, it turned out to be a great disappointment. It was so cloudy that nothing could be seen. Certainly, her disappointment was understandable, for most of us are looking for stars and constellations, not the darkness in the dark sky. "Nothing, just darkness", she said. But what intrigues me is her use of the word "darkness". Quite often in the English-speaking world, when we say "dark sky", we refer to the stars and constellations in the sky, not its darkness; darkness is only the canvas on which they are painted. We use the term "dark sky" often, but what do we understand by its meaning? How does the term "dark sky" work in different cultural contexts? This chapter examines the meaning of "dark sky". Drawing on critical discourse analysis (Fairclough, 1995) and utilising Google Trends data, I examine how "dark sky" is used in English- and Chinese-speaking cultural contexts. Hong Kong is chosen as the location of the case study for the Chinese-speaking context because, firstly, English and Chinese are the two official languages used there and, secondly, Hong Kong is afflicted by light pollution; the problem is so severe that Hong Kong's city light can be easily seen from the International Space Station in space (HKU Light Pollution Research, 2022). A study of Hong Kong will thus offer a chance to explore how "dark sky" is being used in a light pollution-stricken region. This chapter will then provide reflections on the meanings and use of the term "dark sky" in two different cultural contexts.

The meaning of dark sky in the English-speaking context

In English, "dark sky" is a compound composed of one adjective and a noun, with "dark" referring to (1) the absence of light, the black colour (Merriam Webster dictionary online), or (2) night, "of the night or a part of the night" (Oxford English Dictionary online, hereafter OED). In other words, "dark sky" is a compound phrase emphasising the characteristics of the night sky, primarily the absence of light. Historically, the term "dark", as used in "dark sky" was first used in 1953 in the *Sky and Telescope* to discuss the conditions in which a meteor could be observed (Lovell, 1953, p. 122). The use of the term is strongly related to whether

DOI: 10.4324/9781003408444-18
This chapter has been made available under a CC-BY-NC-ND 4.0 license

the sky is "dark" enough for astronomical observations. Due to this historical use and context, "dark sky" differs from its counterpart "night sky", which may not necessarily be "dark" (enough for stargazing). The OED definition, to "dark sky" as "designating a location or set of conditions in which the darkness of the night sky is relatively free of interference from artificial light; of or relating to a night sky having this quality, especially when suitable for astronomical observation". In other words, "dark sky" refers to a specific "area", a location, and/or a specific condition. "Dark sky" is geographically and visually specific, a type of night sky, while "night sky" is a general, temporal state which may not be dark; the two terms are not synonymous.

Although "dark sky" may first have been used in 1953, the phenomenon of the night sky not being dark emerged in 18th-century Europe. With modernisation and urbanisation and the development and increasing use of lighting technology, the day could be lengthened "to every man's fancy" (Bowers, 1998, p. 3). This can be evinced by how, in the 18th and 19th centuries, the upper class used the late hours to demonstrate their honorary social status:

> Two hundred years ago, Parisians ate their main meal [dinner] at 12 o'clock midday; today, the artisan eats at 2 p.m., the merchant at 3 and the clerk at 4; the nouveau riche, entrepreneurs, and bill brokers at 5; Ministers, Deputies, and rich bachelors at 6.
>
> (Schivelbusch, 1988, p. 140)

Not only were meals and schedules altered, but other timeframes for urban practices were also increasingly occurring at later hours: theatre performances began at 7:00 p.m., finishing at 9:00 p.m., after which there might be a visit to a casino or a venue for a post-theatre supper (Schivelbusch, 1988). Such transformed cultural practices were not only limited to the leisured middle and upper classes; as commercialisation of the night-time grew, some shops remained open for diverse customers until midnight. With more and more artificial light being produced by urban areas, the issue of light pollution emerged and began to impact the night sky. Although it is beyond the scope of this chapter to study every use of "dark sky" in the 21st century, its general usage can be illustrated by a small text-mining study. Using Google Trends, Table 13.1 demonstrates the search results for "dark sky" yearly from 2004 to 2022.

The "related queries"—the most common phrases and keywords sought together with "dark sky" through Google reveal two important pieces of information related to the discussion: topics and collocations. These search results notify us about the topics related to "dark sky" through the decades. Summarised here, the search is related to two different topics: (1) the name of a film production house (Dark Sky Film), novel or branded commodity, musical act, or album or song (41%); (2) designated dark sky reserves and light pollution (59%). The results show that in an English-speaking context, "dark sky" is related to either cultural production or astronomical designations.

Table 13.1 Google Trends search results, with the keyword "dark sky".

	Region	Related queries	Note
2004	Iran	International Dark-Sky Association	
	Indonesia		
	Bulgaria		
2005	Taiwan	Dark sky society	
	Canada	Dark sky film	Dark sky films (horror film production and distribution company)
	Switzerland		
2006	Canada	*Texas Chainsaw Massacre*	
	United States	Dark sky films	
	Brazil	IDA	Referring to designation of Brazilian dark sky park
2007	Bangladesh	Dark sky Scotland	
	Serbia	International Dark-Sky Association	
	Ecuador	Dark sky lighting	
2008	Puerto Rico	Dark sky complaint	Uli Jon Roth
	Chechnya	Uli Jon Roth	(German guitarist)
	United States	Dark sky film	*Under a Dark Sky* (music album)
2009	Hungary	Kichler lighting	
	NZ	Dark sky park	
	Canada	Dark sky map	
2010	Belarus	Higher than dark sky	Song by J-pop band *Access*
	Serbia	Access higher than dark sky	
	HK	Dark sky sites	Meteor shower
2011	Puerto Rico	Dark Sky, Radius	Dark Sky (band name) and EP; "Neon", song by Dark Sky; "Brave", song by
	Sri Lanka	Dark sky neon	Kelis (Dark Sky remix)
			The Witcher 2 (fantasy action film)
	UK	Dark sky app	
2012	New Zealand	Dark Sky: "Black Rainbows"	Song
	UK	Milky Way	
	US	Swatch Dark Sky	Brand of watch designated by colour range of face
2013	UK	Jasper Dark Sky Festival	
	Canada	Dark Sky: "In Brackets"	Song
	US	Northumberland dark sky	
2014	Canada	Dark sky imaging	Advice on nocturnal photography
	UK	Kerry Dark Sky Reserve	
	Puerto Rico	Dark Sky: "Rainkist"	Song
2015	S Africa	Big Sean: "Dark Sky (Skyscrapers)"	Song
	US	Big Sean	
	Canada	Big Sean	

(Continued)

Table 13.1 (Continued)

	Region	Related queries	Note
2016	UK	Enya: *Dark Sky Island*	Music album
	Canada	Dark sky festival 2016	
	Ireland	Subaru dark sky	Promotion for dark skies by car manufacturer
2017	NZ	Dark Sky API	Weather app
	UK	Does the sky get dark during a solar eclipse	
	Canada	Why is the sky so dark today	
2018	Canada	Nissan dark sky	Pick-up truck devised to negotiate dark space and containing equipment for stargazing
	NZ	Dark sky metallic GMC	Black brand colour for Ford Sierra car
	US	Jasper dark sky preserve	
2019	NZ	Dark Sky Project, Tekapo	
	US	Massacre Rim, Dark Sky Sanctuary	
	Ireland	River Murray, Dark Sky Reserve	
2020	NZ	Beijing dark sky	Beijing sky goes dark after being hit by a rainstorm
	US	Beijing	
	Ireland	Dark Sky Alternative weather app	
2021	NZ	Faudio Dark Sky	Brand of headphones C J Box (US author of Dark Sky, paperback thriller novel)
	US	C J Box (author)	
	Canada	Dark Sky C J Box	
2022	NZ	Dean Koontz (author)	The Big Dark Sky, futuristic thriller novel
	US	Dark sky reserve India	
	Canada	The Big Dark Sky, Dean Koontz	

The use of "dark sky" as the name of a film production house or cultural output is not my focus here, I now study the second, more dominant topic. Revealed in the Google Trends' search results are the common collocations related to the second topic, which include "dark sky reserve", "dark sky preserve", "dark sky complaint", "dark sky park", "dark sky festival", and "dark sky projects". On a denotative level, these collocations seem to centre on "dark sky": an area reserved for a dark sky, a complaint about (not) having a dark sky, a park featuring a dark sky, and a festival or project about the dark sky. Yet as Google Trends cannot deliver more than these keywords, it is necessary to look further to examine the usage and the meanings of the terms. In what follows, an examination of dark-sky-related organisations and projects, where these terms are used, introduced, and reproduced, will be conducted.

"Dark sky" in dark-sky-related organisations and projects

The collocations suggested by Google Trends seem to suggest that the discussions should centre on "dark sky". However, a closer study of DarkSky (formerly the International Dark-Sky Association), the most important organisation in promoting and preserving dark sky, and one of the key search results in Google Trends in 2004, suggests otherwise. As stated on their website, the vision and values of DarkSky aims to protect the darkness of the night sky; the organisation regards the night sky as a piece of cultural heritage that gives rise to different cultures and traditions: "The night sky, filled with stars, is celebrated and protected around the world as a shared heritage benefitting all living things" (DarkSky, 2023a) It is the star-filled sky, not the darkness, which is the centre of the focus. This can be further illustrated by an activity organised by DarkSky. As misleading as the name of the activity suggests, Capture the Dark Photo Contest (DarkSky, 2023b), it is never the aim of the contest to literally capture the darkness but to bring out the main "actors". It is the stars and constellations that should be captured in the photo. Likewise, another organisation in New Zealand, the Dark Sky Project (DSP), aims to preserve the dark sky and make the night sky "accessible to all" and "connect our Manuhiri (visitors) to the night skies" (DSP, n.d.). However, what is most crucial is "what lies above", that is, the stars. In another activity organised by the DSP, the "dark sky experience", once more the aim is to learn about Māori astronomy than to literally experience the "darkness" (ibid.). Many similar examples can be found in other organisations that refer "dark sky" in their projects. For example, Canada's Jasper Dark Sky Festival, foregrounds and several British dark sky festivals focus on stargazing activities (Dark Skies Festival, n.d.). "Dark sky reserve" and "dark sky preserve" fully reveal the entailment: the terms refer to a specific status wherein the dark sky is reserved for the purposes of stargazing. Similarly, the term "dark sky discovery" typically refers to what can be "discovered" in the dark sky, not the darkness, per se (Dark Sky Discovery, n.d.). As noted by the Council for the Protection of Rural England (CPRE), "[W]e believe that being able to gaze at a starry sky is one of the things that make our countryside so special, and we're working to make sure that we can all experience truly dark night skies" (CPRE, 2022). The reason that "dark sky" instead of "starry sky" or "star-filled sky" is used in organisations' and projects' names might be because darkness is endangered in most of the urban areas in which the majority of the populace reside. The term "dark sky" is thus used to signify the essential conditions for stars to be visible. Further, it is a metonymy that may further enchant the mysteriousness of stars: the sense of origin and nature of the stars and sky. Thus, "dark sky" is not really about the darkness in its reproductions and applications; it works on a connotative level. Most often, it refers more to stars and the star-filled sky. Darkness itself is not the object of appreciation.

An important question therefore arises: what does "dark sky" mean in the Chinese-speaking context of Hong Kong, where light pollution is the worst on the planet (Shadbolt, 2013)? Accordingly, I carried out a similar web search that focused exclusively on links from within a Hong Kong context.

Table 13.2 Google Trends search results, with the keywords "dark sky", "night sky", and "starry sky" in Chinese in the Hong Kong region, 2004–2022.

Keyword	Search results
"Dark sky" in Chinese	No search result
"Night sky" in Chinese	"Star" (星星)
	"The brightest star in the night sky" (夜空中最亮的星)
	"Starry sky" (星空)
"Starry sky" in Chinese	"Swallowed star" (吞噬星空)
	"Gamesky" (遊民星空)
	"Star Trek" (星空奇遇)

In Chinese, there is no equivalent collocation to "dark sky". "Dark sky" (漆黑的夜空) is a possible and grammatically sound combination in Chinese, but it is rarely used. The closest translation would be "night sky" (夜空). This search in Google Trends illustrates this pattern of language use. Whereas *no* search result is found with "dark sky" in Chinese (漆黑的夜空/ 漆黑夜空), some results are shown with "night sky" (夜空), reflecting that this is the common collocation used in Hong Kong. Topping the search result list of "night sky" is "star" (星星), followed by "the brightest star in the night sky" (夜空中最亮的星) and "starry sky" (星空). This evinces that "dark sky" in English can be roughly translated as "night sky" in Chinese and, more specifically, in the Hong Kong Chinese-speaking context.

Most worthy of note are the Google Trends search results for "night sky" in Chinese, which without exception, are related to music referring either to the titles or lyrics of pop songs. Certainly, the top search result "star" can also denote any stargazing-related activity, but these possible double significations do not undermine its strong relationship to cultural representations. Another search of "starry sky"/"star-filled sky" demonstrates that stars and their strong correlation with popular culture is not a simple coincidence, for it produced the following results. Most numerous is 吞噬星空 (*The Legend of the Spacewalker*), the second is 遊民星空 (*Gamersky*), and third most popular link it to 星空奇遇 (*Star Trek*). Unlike the search results in the English-speaking context, where astronomical activities are the most dominant result, these all relate to cultural productions, respectively a Chinese science fiction graphic novel and subsequent movie, a video game website, and globally broadcast, seminal TV drama series. They are all specific pronouns, and none focuses on stargazing or constellations. This is not to suggest that astronomical activities such as stargazing are completely absent in Hong Kong, but no astronomically related activities were found in the Google Trends search results. Findings suggest that "star" is most frequently linked with regard to cultural productions because popular representations occlude in popularity any reference to marginal astronomical pursuits.

Despite a Google Trends search producing no results, it is still possible that the term "dark sky" is being used in Hong Kong. Accordingly, I explored whether a more focused search across Hong Kong's four most popular Chinese language newspapers, *Ming Pao News*, *Oriental Daily*, *Singtao Daily*, and *AM730*, might result in more references. The search for "dark sky" from 18 August 2021 to 17 August 2022 resulted in the discovery of only five references. Four out of the five results for the search "dark sky" are *not* related to any astronomical activity: two discuss the Chinese spaceship and its debris (Oriental Daily, 2022a; Sing Tao Daily, 2021a), one talks about how fireworks left traces in the dark sky during the Para-lympics in 1992 (Chan, 2021), another discusses an author having a workout on the hot summer night (Wat, 2022), and the last one is a discussion about bats at night (Sing Tao Daily, 2021b). However, a parallel search for "night sky" produced 138 results. In comparison, "night sky" is used in many different scenarios, including a discussion on fireworks (Mingpao, 2022a), mid-autumn festival (Wong, 2021), the night sky in the rural area (Sing Tao Daily, 2022), and stars and astronomical discussions, such as the starry night sky in ancient China (Tsui, 2022), the visibility of Jupiter, and the super full moon. These two search results further confirm my suspicion that there is no equivalent to the English term "dark sky" in the Hong Kong Chinese-speaking context. In Hong Kong, "dark sky" is rarely used, a term related to its literal, denotative meaning of "darkness" than "stars". "Night sky", on the other hand, is the term more widely used, especially when astronomical activi-ties are involved. In short, in the English-speaking context, "dark sky" is related to stargazing and astronomy-related activities, whereas "night sky" is related to these practices in the Hong Kong Chinese-speaking context. Why might this be so? If language reflects the condition and culture of a place, it suggests that in Hong Kong there is a night sky but it is not a dark sky.

Hong Kong residents are very accustomed to a bright night sky, a sky so bright that darkness has almost disappeared. It is reported that in the urban areas where light pollution is most serious, the night sky is 1,200 times brighter than a sky with-out light pollution (Pun *et al.*, 2014). Not even rural areas are spared, with the night sky in the Wetland Park having been recorded at 130 times higher than the stand-ard, with the skies lit up by nearby urban areas (ibid.). The light pollution problem has persisted, if not worsened, in the 2020s. Although a charter of light pollution measures was issued by the Hong Kong government in 2016, light pollution com-plaints have continued to increase from 234 in 2011 to 373 in 2020 (Sky Post Daily, 2022), and the night skies in the urban areas remain almost nine times brighter than the international standard (The Green Earth, 2022). (See also Figure 13.1.)

Though night sky is not dark in Hong Kong, many do not see this as a problem. A dark sky is necessary for astronomy lovers but appears not be a major concern for the general public and the Hong Kong government, which narrowly defines light pol-lution as a problem of energy wastage (Lam, 2022). Moreover, for some, the notion of the dark night carries negative meanings. For instance, for the commercial sector, external lighting beautifies Hong Kong the Pearl of Orient (Task Force on External Lighting, 2013). For these people, the attractiveness of Hong Kong lies in its energy and vibrancy, fostered and catalysed by artificial lights. Darkness, in this context, means lifelessness and being "out of business". As stated clearly in a government report, "the darker environment would drive the already deteriorating business envi-ronment at night further downhill" (Task Force on External Lighting, 2013). Further, the government also holds that it is vital to "maintain Hong Kong's famous night

Figure 13.1   Night or day? Urban areas are lit up at night in Hong Kong. Photo by Shine
Shine Law (2022).

scene" (Advisory Council on the Environment, 2011); "it would not be realistic to
expect light nuisance to be eliminated in Hong Kong" (ibid.). Being proud of one
of the "Top 3 Night Views in the world" (Mingpao, 2012), Hong Kong chooses to
embrace the artificial night view to define and demonstrate its identity; the use of
light is regarded as a Hong Kong signature. "Dark night" or "darkness", in this specific context, connotes a sense of sluggishness. The brightly illuminated sky in Hong
Kong is well-accepted, even celebrated, and this is undergirded by how darkness
connotes a sense of negativity, signifying the unknown, death, and in Greek mythology, another world (Aguirre, 2010) or sense of otherness (Karakantza, 2010). We are
afraid of darkness (Klinkenborg, 2008). Similar connotations are prevalent in Hong
Kong: in news reports, dark or darkness is used to use as a simile for the bad times,
for crime, and triad societies (Oriental Daily, 2022b), for economic hardship (Oriental Daily, 2022c), poverty (Mingpao, 2018) and for a generally bad and difficult time
filled with social problems (Oriental Daily, 2023). Therefore, in Hong Kong, it is not
surprising that the dissipation of the dark may be considered good news.

This is not to suggest that astronomical activity is entirely absent in Hong
Kong (Law, 2020). Quite often, when there is an event such as a meteor shower,
the importance of darkness might still be referred to in the news. Nevertheless,
astronomical activities are deemed rather insignificant, as illustrated by the 2013
government report on light pollution, in which some of the complainants are

referred to as "light-sensitive receivers" (Task Force on External Lighting, 2013). Furthermore, light pollution is only defined as a problem of energy wastage and light as a nuisance, where the importance of the natural dark sky is simply absent from the discussion (ibid.). The same definition can also be found in other reports, where, for example, a regional study of light pollution fails to even mention both astronomical activity and the natural dark sky (HKBU and CASR, 2016). And in a news report concerning light pollution in 2021, light as a nuisance remains the dominant narrative (Oriental Daily, 2022d). I argue that, in order to maintain Hong Kong's unique identity, the government and the community have chosen to embrace artificial lights, rather than starlight, or the dark sky. Stargazing is regarded as less important and less glamorous by many, as shown in Figure 13.2, where a bright moon is ignored in favour of the artificial lights arranged "artistically" to mimic nature.

We could argue that the Hong Kong public has forgotten the dark sky. This can be showcased by an unexpected power outage incident in the district of Yuen Long, on 21 June 2022. The district's residents were left without electricity until power resumed on 23 June. Drawing on *Wisenews*, a news search engine, we can glimpse the activities and concerns of the residents and the public regarding the incident, with residents' experiencing discomfort and inconvenience (Mingpao, 2022b), the commercial sector struggling to run their businesses during the blackout, and the creative uses of light in the households of Yuen Long, where, for example, residents used candles and lightboards produced for pop concerts to provide lighting (AM730, 2022). What is of concern here is that stargazing or enjoyment of the dark sky during these power outrage periods was seemingly absent from the discussions. Unlike the picture books *Bright Sky, Starry City* (Krishnaswami, 2015), where the residents rush out to appreciate the dark star-filled sky during a power outage, and *The Fairy who Loves Beauty*, which features nocturnal animals (Mak, 2012), the Yuen Long residents appear to neglect, or simply forget, the possibility of gazing at the dark sky during the power outage. The lights remained the major emphasis, with residents forgetting all about the stars amid the darkness. The language used is a simple reflection of the situation in Hong Kong.

Stars and constellations are seldom a regular subject in cultural representations in the Hong Kong Chinese-speaking community. Although stars and starry sky are frequently mentioned in song titles, almost without exception, the actual star is not the primary subject in the lyrics; stars exist only as the setting for the unveiling of the human love story, with the night sky only occasionally referenced, and darkness rarely. Some ancient Chinese poetry reveals a similar attitude, as exemplified by a poem that many Chinese know by heart:

"Quiet Night" (靜夜思), by Li Bi (李白)

床前明月光，	Beside my bed, the bright moonlight,
疑是地上霜。	I mistook it for frost on the floor.
舉頭望明月，	Lifting my head, I gaze upon the bright moon,
低頭思故鄉。	Lowering my head, I think of my hometown.

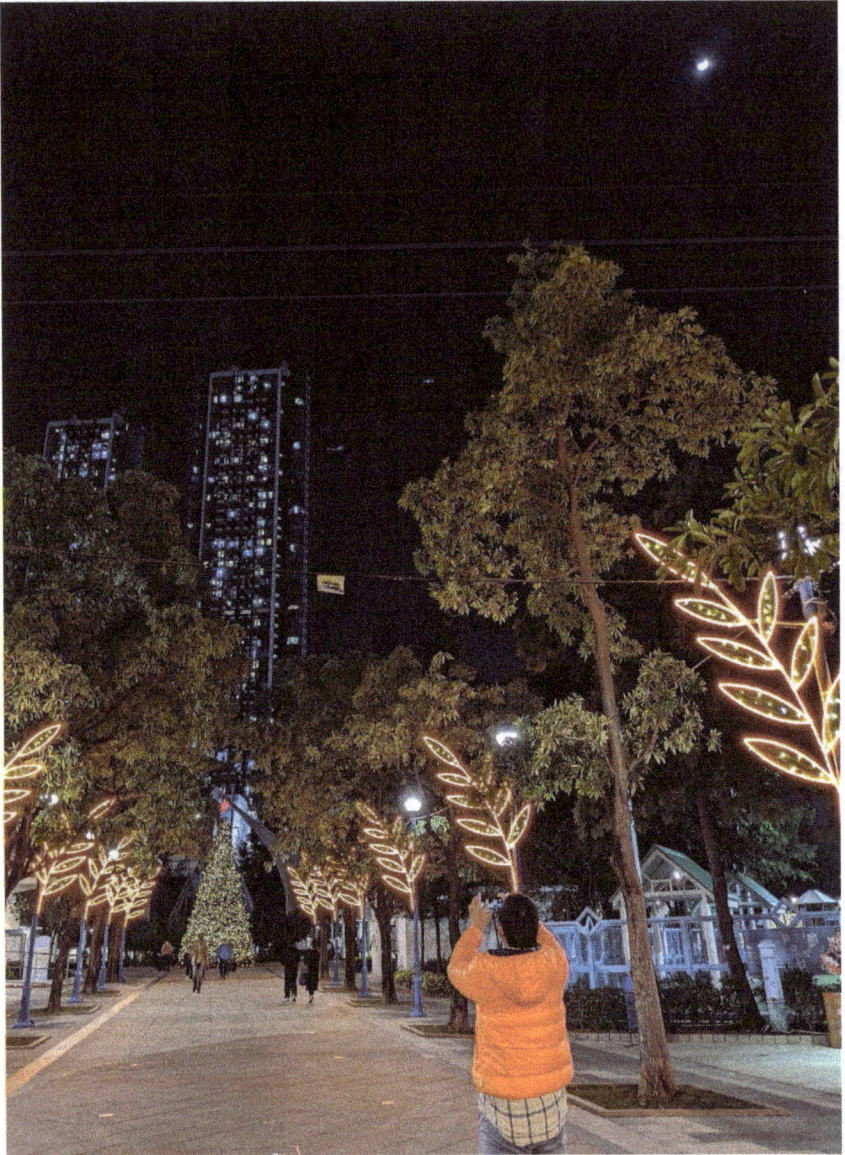

Figure 13.2 A resident focuses on the artificial light decorations, ignoring the bright moon
high above. Photo taken by the author.

In this Chinese Tang poem, the author's homesickness, not the moon, is the
major motif. Further, note the time of the event in the poem. When did it take
place? It was night-time. How is this signified? Through the presence of the moon.
This is the usual way in which "night" or "darkness" is denoted in ancient Chinese
poetry. As evinced in another very well-known poem "Short Poem Style" (短歌

行), by Cao Cao (曹操), the famous line "Bright moon sparse star (月明星稀), South did the magpies fly (烏鵲南飛)" describes bird migration at night, with the night-time connoted by the presence of the bright moon. As we can see, the words "night" or "darkness" may not be used to illustrate time. I therefore contend that stars are seldom the major motif in these cultural representations but are vehicles for connoting the night, and darkness is not specified but only implied by the presence of the moon and stars. Perhaps we could even say that neither darkness nor the starry sky matter by themselves in such cultural representations.

The star-less dark sky

In the English-speaking world, "dark sky" works on a connotative level; it does not suggest darkness literally, but it refers to stars. In comparison, in the Chinese-speaking context in Hong Kong, a Chinese term equivalent to "dark sky" does not exist; the terms "dark sky" and "starry sky" are not widely used in Hong Kong; a closer translation of "dark sky" would be "night sky". Implied in the language used is the situation in Hong Kong: there is only a night sky, which is neither dark nor starry and which is quite well-accepted.

As there are already many discussions on the importance of stars (for example, Francis, 2019; Bogard, 2008, 2021; Marlin, 2021), I shall conclude by addressing the darkness in the sky. Darkness, as shown in the previous sections, is only a background for the manifestations of stars in the English-speaking world, important only because of its instrumental value. Unvalued, darkness has been expelled in Hong Kong. I therefore propose the following question: Is it possible to appreciate the darkness as much as the stars? In a conversation with the astrophysicist Dr Wong Wing Hung, he explained the signification of "darkness". He argued that "black" is not an appropriate term to describe the night sky because the term "black" would literally mean black, something we humans have identified; something known. However, "dark" may refer to things yet to be known, including dark energy and dark matter, the unknowns (we know we do not know), and other cosmic phenomena, the unknown unknowns (we don't know we don't know) (Suto, 2010). Darkness, in this sense, is a representation of the mystery of the universe, a recognition of our limited knowledge of the world. It invites our exploration and imagination. Perhaps, we should pay a little more attention to the darkness in the dark sky too.

References

Advisory Council on the Environment. (2011) *External lighting in HK (ACE Paper 4/2011)*. Hong Kong: Hong Kong Special Administrative Region.

Aguirre, M. (2010) 'Erinyes as creatures of darkness', in Christopoulos, M., Karakantza, E. D. and Levaniouk, O. (eds.) *Light and darkness in ancient Greek myth and religion*. Lanham: Lexington Books.

AM730. (2022) 'Yuen long power outage, pop star's lightboards brighten us all up' (元朗停電 靠姜濤燈牌照亮全屋 鏡粉：全靠佢照住)., 21 June.

Bogard, P. (2008) *Let there be night: Testimony on behalf of the dark*. Reno, NV: University of Nevada Press.

Bogard, P. (2021) *To know a starry night*. Reno, NV: University of Nevada Press.

Bowers, B. (1998) *Lengthening the day*. Oxford: Oxford University Press.

Chan (陳耀華). (2021) 'Ryuichi Sakamoto and the Olympics' (坂本龍一與奧運) *Ming Pao*(明報), 7 September.

CPRE. (2022) 'Night skies outlook is bright, our star count results suggest', 17 May. Available at: www.cpre.org.uk/news/night-skies-outlook-is-bright-our-star-count-results-suggest/ (Accessed: 1 May 2023).

Dark Skies Festival. (n.d.) Available at: www.darkskiesnationalparks.org.uk/ (Accessed: 1 May 2023).

Dark Sky Discovery. (n.d.) Available at: www.darkskydiscovery.org.uk (Accessed: 1 May 2023).

DarkSky. (2023a) *Vision and values*. Available at: https://darksky.org/news/vision-values-update/ (Accessed: 1 August 2023).

DarkSky. (2023b) *Capture the dark photo contest*. Available at: https://darksky.org/what-we-do/events/photo-contest/ (Accessed: 1 August 2023).

DSP. (n.d.) *Dark sky project*. Available at: www.darkskyproject.co.nz (Accessed: 1 May 2023).

Fairclough, N. (1995) *Critical discourse analysis: Papers in the critical study of language*. New York: Longman.

Francis, T. (2019) *Dark skies: A journey into the wild night*. London: Bloomsbury Wildlife.

HKBU and CASR (Hong Kong Baptist University and Centre for The Advancement of social sciences research). (2016) *Billboard-causing light pollution in Wai Chan district: An investigation* (灣仔區內招牌引致光污染的調查), Hong Kong: The Wanchai District Office of Home Affairs Department and Community Building Committee of Wanchai District Council (2012至2015年度灣仔區議會社區建設委員會屬下大廈管理工作小組).

HKU Light Pollution Research. (2022) The seriousness of light pollution in Hong Kong. https://nightsky.physics.hku.hk/en-hk/light-pollution-resources/educational-resources/seriousness-light-pollution-hong-kong (Accessed: 21 September 2023).

Karakantza, E. D. (2010) 'Dark skin and dark deeds danaids and aigyptioi in a culture of light', in Christopoulos, M., Karakantza, E. D. and Levaniouk, O. (eds.) *Light and darkness in ancient Greek myth and religion*. Lanham: Lexington Books.

Klinkenborg, V. (2008) 'Our vanishing night: Most city skies have become virtually empty of stars', *National Geographic*, 214(5) (November).

Krishnaswami, U. (2015) *Bright sky, starry city*. Toronto, ON: Groundwood Books.

Lam, Y-M. (2022) 'A light pollution discourse in Hong Kong', *Visual Studies*. https://doi.org/10.1080/1472586X.2022.2060302.

Law (羅保熙). (2020) 'Dark skies should be immediately preserved' (保育星空不應蹉跎歲月 仿韓日台建「暗空公園」), *HK01*.

Lovell, A. C. B. (1953) 'Radio astronomy at Jodrell bank', *Sky & Telescope*, 3, p. 122.

Mak, H. F. (麥曉帆). (2012) *The fairy who loves beauty* (愛美麗的花仙子) Hong Kong: Sun Ya Publications.

Marlin, M. (2021) *Astrotourism: Star gazers, eclipse chasers, and the dark sky movement*. New York, NY: Business Expert Press.

Merriam-Webster. Available at: www.merriam-webster.com/dictionary/dark.

Mingpao (明報). (2012) 'HK night view continues to be one of the best on the list' (香港蟬聯世界三大夜景之一), 6 October.

Mingpao (明報). (2018) 'Future city: How would a lights-off Nathan road be read?' (未來城市：燈飾消失了 彌敦道怎麼說), 31 December.

Mingpao (明報). (2022a) 'A special edition of A Symphony of Lights' will be released in July' (特別版 幻彩詠香江 7月每日上映), 29 June.

Mingpao (明報). (2022b) 'Disconnected at home; Reconnected in the metro station' (家中斷網 市民留港鐵站上網), 22 June.

Oriental Daily (東方日報). (2022a) 'Chinese rocket returning to Earth; Keep away from the rocket debris, the Malaysian gov has urged' (華火箭墜回大氣層 馬國籲勿檢殘骸), 1 August.

Oriental Daily (東方日報). (2022b) 'The more the HK police work, the darker the situation' (愈緝愈毒愈反愈黑 香港沉淪警察運吉), 5 April.

Oriental Daily (東方日報). (2022c) 'Beauty shop folded without warning, customers filing complaints' (美甲店東突結業走佬 眾顧客聲討), 28 April.

Oriental Daily (東方日報). (2022d) 'Voluntary Charter is just a toothless tiger, residents deeply-troubled by external lightings' (光害約章冇牙力 累街坊影響作息), 28 November

Oriental Daily (東方日報). (2023) 'Tax refund + Shopping voucher, Legislative council members urge the Government to take measures to rebuild the economy' (議員促振經濟 推組合拳 退稅 免息貸款 派消費券), 25 January

Pun, C. S. J., So, C., Leung, W. Y. and Wong, C. F. (2014) 'Contributions of artificial lighting sources on light pollution in Hong Kong measured through a night sky brightness monitoring network', *Journal of Quantitative Spectroscopy and Radiative Transfer*, 139, pp. 90–108.

Schivelbusch, W. (1988) *Disenchanted night: The industrialization of light in the nineteenth century*. Translated by Davie, A. Berkeley, CA: University of California Press.

Shadbolt, P. (2013) 'Hong Kong's light pollution "worst in the world"', *CNN*, 21 March.

Sing Tao Daily (星島日報). (2021a) 'Shenzhou entering Tianhe, European space agency congratulates' (神舟三傑進駐天和 歐航天局發文祝賀), 17 October.

Sing Tao Daily (星島日報). (2021b) 'Bat and luck' (有「蝠」氣), 7 December.

Sing Tao Daily (星島日報). (2022) 'In-depth travel, not something trivial' (帶動深度遊 不是趕鴨仔), 2 May.

Sky Post Daily (晴報). (2022) 'Light pollution worst in Mongkok and Tsim Sha Tsui, 9 times higher than the international standard' (旺角尖沙咀光污染重災 平均亮度超國際標準9倍), 25 February.

Suto, Y. (2010) 'Unknowns and unknown unknowns: From dark sky to dark matter and dark energy', Proceedings Volume 7733, Ground-Based and Airborne Telescopes III, p. 773302.

Task Force on External Lighting. (2013) *Document for engaging stakeholders and the public*. Hong Kong: Hong Kong Special Administrative Region.

The Green Earth (綠惜地球). (2022) 'Light pollution is 9 times higher than the international standard. (香港光污染重災區約為國際標準9倍 問題嚴峻受國際關注)', *Press Release*, 24 February.

Tsui (徐瑋). (2022) 'When the ancient treasure meets the HK artist' (古今無界當故宮國寶遇上香港藝術家). *Mingpao* (明報), 13 August.

Wat (屈曉彤). (2022) 'Midsummer' (仲夏), *Ming Pao* (明報), 28 July.

Wong (黃國軒). (2021) 'Working as hard as the stars' (像星星一樣努力發光). *Sing Tao Daily* (星島日報), 15 September.

Part 6

Dark sky tourism

14 Tread softly in the dark

Georgia MacMillan, Hannah Dalgleish,
Therese Conway, and Marie Mahon

Introduction

On 14 February 1990, NASA spacecraft Voyager 1 captured one of the most significant photographs ever taken. It was part of a series of photographs of the solar system known as the *Family Portrait* and featured a remote image of the Earth pictured as a fragile pale blue dot suspended in a glimmer of light, in the vast cosmos.

This image of Earth, as viewed from four billion miles away, is considered to be one giant "selfie", instigated by humans, of humankind. The accompanying words of Carl Sagan (1994), American astronomer and science presenter, directed a collective gaze upon the world:

> Look again at that dot. That's here. That's home. That's us. On it everyone you love, everyone you know, everyone you ever heard of, every human being who ever was, lived out their lives.

Thirty years on, low-orbit space tourism has become a reality and sustainable tourism is under critical analysis (Frost and Frost, 2021). The image of the pale blue dot serves as a stark reminder of the contrasting nature of tourism and its ethical responsibility to the planet and to the living creatures residing on it. As the tourism industry moves into a post-pandemic era, recovery frameworks designed to revive the industry and assist with its resilience are central considerations (Daddi *et al.*, 2020). This, as well as the resultant focus on developing more "soft", sustainable, and ethical offerings, affords a major opportunity for dark sky tourism (DST) to capitalise on its position as a sustainable form of tourism. Further potential exists to combine DST with circular tourism, and it also posits the opportunity for wellness tourism with many travellers seeking dark sky places as sites that offer "a way to reconnect with the natural world during a time when we spend most of our day interfacing with technology" (Foster, 2017).

DST involves travelling to remote landscapes that have been deemed wild and free from light pollution, partly to experience celestial phenomena (Dalgleish and Bjelajac, 2022), but also to engage in other forms of tourism that lends itself to remote and wild places, such as night hikes, storytelling, and cultural experiences,

DOI: 10.4324/9781003408444-20
This chapter has been made available under a CC-BY-NC-ND 4.0 license

all set within the increasingly rare opportunity to experience natural darkness (Khetrapal and Bhatia, 2022).

In his discussions on ecotourism, Weaver (2011) connects nature with celestial observations, thus leading a visitor's gaze upwards to engage the often-neglected sky (Blair, 2016). Weaver's perspectives focus on the night sky as offering a natural resource for nature-based tourism experience. Blundell, Schaffer and Moyle (2020) apply the theory of last chance tourism (LCT) to DST, identifying that the growth of anthropogenic light pollution is reducing the number of locations where a naturally dark sky can be experienced (outside of dark sky accredited places). The resulting framework views DST through a more holistic lens of ethical sustainability and highlights its links with LCT as well as foregrounding the cultural and ecological impacts of losing the resource of a naturally dark sky.

Worldwide, over 200 dark sky places have been sanctioned by DarkSky (previously known as the International Dark-Sky Association) and other awarding authorities, providing destinations for DST. Dark sky areas are often places of existing conservation, such as national parks, connecting the physical, terrestrial landscape after dark with celestial viewing. Tyler Nordgren, space artist and astronomer of the US National Parks, trademarked the expression "Half the park is after dark" (see Avery, this volume) specifically to highlight the connection with nature and the night. These places offer an extension to the tourism season, providing an extra reason to visit these scenic destinations outside of the traditional tourism period and in darkness.

Dark skies in sustainable tourism

Sustainable tourism is defined by the United Nation's World Tourism Organisation (UNWTO) as "tourism that takes full account of its current and future economic, social and environmental impacts, addressing the needs of visitors, the industry, the environment and host communities" (UNWTO, 2013). It is economically relied upon by many countries from the Global North and Global South and provides a source of growth in employment and diversification across many sectors. In the context of sustainable tourism, the United Nations' Sustainable Development Goals (SDGs) are strongly debated, particularly in relation to reducing socio-economic inequalities, enhancing wellbeing, and including traditional ecological knowledge to empower local communities (Boluk, Cavaliere and Higgins-Desbiolles, 2019). DST has the potential to address some of these shortcomings and demonstrates a synergy with many of the SDGs such as wellbeing, community engagement, climate change, biodiversity, and lifelong learning (Dalgleish *et al.*, 2021). As an example, Jacobs, Du Preez and Fairer-Wessels (2020) explore the market in South Africa, with positive findings about the contribution of DST to off-season tourism product diversification, job creation, and economic prosperity. Furthermore, DST may also produce a lower carbon footprint in comparison to other forms of tourism and further posits an alignment with the concept of degrowth in tourism (Gajdošíková, 2023). For example, DST offers diverse opportunities to (1) save energy and preserve ecosystems by reducing artificial light at night; (2) embrace slower tourism and empower locals by extending the duration of overnight stays in a location; (3) preferentially attract domestic tourists who do not need to fly; (4) rebalance demand, spreading the load from high-peak summer times toward off-peak times when the nights are longer; (5) educate tourists on light pollution, which could lead to behavioural changes and inspire them to advocate for darker skies within their local communities; and (6) connect with complementary themes such as mindfulness and wellness tourism.

Therapeutic nightscapes

The inclusion of wellbeing and mindfulness activities at dark sky locations also draws health and wellness tourism into the offerings provided by DST. Retreats featuring darkness therapy, moonlight yoga, meditative experiences in darkness, and guided nature walks at night are activities easily transferred from existing health tourism models. Health benefits derived from time spent in nature are generally associated with daytime experiences described as taking place in green and blue spaces, namely, in water bodies or natural landscapes (Foley and Garrido-Cumbrera, 2021; Brewer-Smyth, 2022). Whilst these colourful daytime references, part of what has become known as palettes of place, describe therapeutic landscapes, health geographers have acknowledged the limitations which overlook more embodied, sensory experiences (Bell *et al.*, 2018). The kind of immersive practices discussed by Edensor (2013) suggest there are opportunities to be explored that are generated from reconnecting and immersing oneself in darker spaces, attaching meaningful awe-invoking and potentially spiritual experiences.

The concept of awe, a powerful emotional response to a phenomenon, is also mentioned in several discussions on night sky observations (Blair, 2018; Marr, 2020). Drawing on a multidisciplinary social science approach, Derrien and Stokowski (2020) describe how individuals attached multiple meanings to the night sky, ranging from personal associations based on prior experiences to feelings of awe. Others demonstrate that total immersion in darkness can be experienced in dark sky places, wherein the seamless connection of land and sky without a visible horizon produces feelings of being embodied and suspended by the sky itself (Blair, 2016). This suggests a connection to our place in space, without the need to travel beyond *terra firma*. Exploring darkness requires a visual adaptation whilst allowing other senses of sound, touch, and smell to define the environment and heightening scotopic sensitivity to shadow and contrast (Dunn and Edensor, 2020). Johan Eklöf invites those unfamiliar with these senses to *carpe noctem* and seek out the darkness to experience the life-enriching rhythm of day and night (2022). Indeed, a therapeutic nightscape can be consumed on the doorstep as a new sensory exploration, while satisfying the goals for sustainability and drawing novel qualities of wellbeing and mindfulness into sustainable tourism experiences in a holistic way.

Cultural connections with the night sky

Over millennia, the human perception of a starry night has attributed cultural meanings, personalities, stories of heroic deeds, myths, and legends to the visual phenomena of a celestial night skyscape. UNESCO acknowledges such cultural connections and the significance of a naturally dark sky in its Portal to the Heritage of Astronomy programme, established 2012 in collaboration with the International Astronomical Union. Further, the night sky and its connection to the belief systems of indigenous people are considered a cultural resource by the United States' National Parks Service (Rogers and Sovick, 2001). Recognising traditional knowledge enables communities to explore their relationships with the night sky and to deliver authentic, place-based approaches to DST. Exploring the influence of the night sky on traditional knowledge systems reflects how close human engagement with a nocturnal world once was, until it became disconnected by the impact of light pollution in what is considered an "extinction of experience" (Davies *et al.*, 2013, p. 1).

To reclaim the natural and cultural resource of a night sky, researchers have called for the end of light pollution on behalf of indigenous communities and the restoration of human relationship with wild, dark nights (Khetrapal and Bhatia, 2022; and see chapter by Khetrapal in this volume). Hamacher, De Napoli, and Mott (2020) go as far as to say that whitening the sky is a form of cultural genocide, while Pereira, Archibald, and Dalgleish (2022) claim that access to a dark, starry night sky is a human right. According to the United Nations (2022, 2018), humans have the right to a "clean, healthy and sustainable environment", as well as the right "of access to and enjoyment of cultural heritage", which reinforces the need to rewild the night sky for both environmental and cultural reasons.

The link between indigenous knowledge and DST is seldom made in the literature. Auala *et al.* (2023) conducted research with indigenous communities in Tsumkwe, Namibia, to explore their relationship with the stars and their perspectives on DST. The community felt that DST offered them opportunities to attract tourists, diversify their activities, and share their knowledge and culture. Yet the authors find that indigenous stories about the stars are endangered; traditional knowledge is not passed on as it once was, and so the starlore in Tsumkwe—which currently remains with only one person—may be lost altogether if not preserved. There may well be a similar case to be made for other cultures and traditions around the world. Bringing awareness of DST to indigenous groups offers a way to retain astronomical folklore in these communities, while demonstrating the critical need to pursue dark sky research and preserve oral histories and cultural heritage before they are lost altogether. This will ensure a future for indigenous peoples as cultural custodians of the night.

Indigenous storytelling has been shown to be important for stakeholder engagement (Binneman and Davis, 2020) and conservation (Fernández-Llamazares and Cabeza, 2018), and so preserving indigenous narratives may have additional benefits for cultural and environmental conservational pursuits. The Astrostays project in the Indian Himalayas is one example of this practice. Founded in 2018,

Astrostays has, at the time of writing, trained 35 women to be astronomy guides in more than 15 different villages in Ladakh. These women educate visitors in astrophysics, local indigenous culture (including starlore), and gastronomy. The community-led project ensures that money is reinvested in local infrastructure, like greenhouses and solar water heaters. Astrostays has also attracted an unexpected tourist demographic (mainstream tourists rather than amateur astronomers); more than 1,200 tourists have visited, bringing in tens of thousands of dollars in revenue. There are socio-economic benefits too. Astrostays provides an incentive for skilled people to remain in their rural communities, helping to stimulate the local economy which further supports the SDGs.

Such deeply intangible connections to the night sky are not as clearly evident in Western societies. However, there are plenty of tangible heritage constructs in archaeology that connect the skyscape with the landscape, or light with darkness, given that many monuments and megalithic art forms were deliberately concealed in shadow (Dowd and Hensey, 2016). The significance of the sky in archaeology is more commonly known as archaeoastronomy and is prominent in the celestial alignments of megalithic tombs across Europe, such as Stonehenge in the UK and Newgrange in Ireland (Hensey, 2021), both listed as UNESCO World Heritage Sites that attract international audiences who travel to experience their winter and summer solstice phenomena. Yet an influx of vast tourist numbers to such hallowed sites causes harm to the environment and the protected site if not managed appropriately, raising questions of sustainability, and prompting custodians to restrict attendance numbers.

Dark skies and community participation

Embracing DST is only possible if host communities value the natural resource of darkness. As an example, Wild Nephin National Park (the newest of Ireland's six national parks), located in County Mayo in the west of Ireland, becomes Mayo Dark Sky Park by night. While it is managed by the National Parks and Wildlife Service, its formation was somewhat unconventional, and its origins lie in the surrounding communities. At first glance, Mayo is an unlikely host for a dark sky place given the late arrival of artificial lighting to the county, with some houses remaining off-grid until as recently as 1976. Mayo also suffered significant periods of emigration, leaving it one of the largest (5,588 km^2) yet least densely populated counties in Ireland (137,231 residents in the 2022 census). Symbolically, light in the form of a burning ember represented the family hopes of return and was traditionally kept alight by the remaining community members (Fitzgerald and Lambkin, 2008).

Places seeking international dark sky accreditation typically evolve from either a tourism initiative or through the concentrated efforts of amateur astronomers. In late 2013, Mayo had neither. Instead, it had a strong base of willing volunteers, some of whom were recent additions to the community (known as "blow-ins"), who formed the Friends of Mayo Dark Skies Group (representing the communities of Newport, Mulranny, and Ballycroy, bordering the National Park). Outsiders are often able to recognise idiosyncrasies in a locality that for locals are unnoticed, and in this case, they saw the darkness that pervaded much of the area. By 2016 the Friends of Mayo Dark Skies had undertaken the journey that culminated in the assignation of dark sky status and a gold tier International Dark Sky Park was awarded to the national park, thereafter to be known as Mayo Dark Sky Park. This sparked the formation of a Newport Astronomy Club, which held regular events, whilst those interested in outdoor pursuits launched Terra Firma Ireland (co-founded by one of the authors), a social enterprise exploring the potential for DST that offered a suite of experiences from the Greenway to the Milky Way. Terra Firma Ireland combines activities that range across storytelling, close-up magic, and stargazing on nights of Magic, Myth, and Moonlight, and the organisation guides small groups on board the Starlight Express as it follows the north star on a dark sky safari. Almost paradoxically, given Mayo is a relative newcomer to astronomy, this cooperative network of community groups and social entrepreneurs that champion the dark sky park has been called to action by a determination to protect the night sky through public engagement. The annual Mayo Dark Sky Festival, organised by the communities and the dark sky park, has become a flagship event of the county, and is no longer considered a niche event for astronomy enthusiasts. Described on its website as a "three-day extravaganza, the festival explores themes of astronomy, cultural heritage, and the environment in a fusion of family-friendly events (Mayo Dark Sky Park, 2022). The eclectic programme attracts internationally renowned speakers and celebrates the surrounding natural nightscapes with science, art and storytelling, rocket-making workshops, and environmental education talks. The festival takes place each November, encouraging visitors to stay more than one night at a time that is outside of the typical tourism season, supporting claims that DST is an apposite vehicle for sustainable tourism (Dalgleish and Bjelajac, 2022).

Other global initiatives such as International Dark Sky Week (IDSW) demonstrate the power of civil society in engaging communities on the journey to dark sky advocacy, enabling them to showcase their potential as a destination for DST. In Mayo, local communities host their own participative events for IDSW through a programme curated to fuse the identity of each community with a dark sky theme. Held annually in springtime, this week of events—the Dark Sky Roadshow of Mayo—marks the seasonal end of darker nights. The following provides two accounts of events staged in 2023 from a practitioners' perspective that have been selected for their contribution to fostering community participation and placemaking through dark sky initiatives.

First, the remote village of Blacksod, home to a landmark lighthouse at the edge of the Atlantic Ocean, designed a participative evening exploring the past practices of night sky navigation used by local mariners (Figure 14.1). Historians and coastguards told the story of the Great Lighthouses of Ireland, their part in saving lives at sea, and the past times during which the sea was a highway of transportation and Ireland's fishing heritage supported this region. Fittingly, the setting for the event was located next to an iconic fishing boat by the harbour wall and attendees,

Figure 14.1 Felicity Egan (2023), *Navigating the Night Sky*, Blacksod Lighthouse, Co. Mayo 2023.

some of whom had travelled hours to be there, gathered with tea and deckchairs. A yoga-practitioner-led interludes of stretches to keep warm and moments of silence for mindful reflection. Participants used this time to remember those who once depended upon this Atlantic coast for their livelihoods and survival and those who had departed this continent from this small harbour, seeking new horizons and a new world. The evening revived multiple connections—connections to the physical environment, such as the sea and this wild shoreline, and social connections to neighbours, which constitute necessary dimensions for wellbeing and community stewardship (Kelly, 2018). Whilst the very idea of a lighthouse, a tourism attraction in its own right, may appear contrary to the ethos of DST, there is a symbiotic duality between the attractions; the iconic beacon of a lighthouse holds mystery and fear to some yet sanctuary and tranquillity to others (Azevedo, 2018).

Second, Ireland is popularly known as "the land of saints and scholars", and the written and spoken word is an important aspect of its rich culture. In another Mayo community, the use of a literary theme through the lens of dark skies led to an innovative evening for IDSW. Local writers and reciters gathered in Ballycroy to take part in *Dánta in the Dark* (*dánta* means "poems" in the Irish language) to celebrate the wonder of night. The writings included personal narratives, local traditions, and concerns for the conservation of the night. Linking the arts to dark skies is not a new concept; many musical lyrics express human curiosity about our celestial ceiling. However, the format of this event prompted unexpected interpretations of what is important to the community about darkness and dark skies, drawing on emotional attachments to nostalgia and pro-environmental values. The poems and musings curated for the event programme were printed as a commemorative booklet and distributed to the audience. The resulting publication provides a physical representation of work by professional and amateur wordsmiths inspired by the dark sky theme. What has emerged from these accounts is the incorporation of local socio-environmental culture in these participative events. Establishing custodianship in this way adds authenticity and firmly places the community as leaders of their own initiatives, an essential component in sustainable tourism (Eslami *et al.*, 2019).

Of equal importance to Mayo's dark sky communities is the active involvement in how their environment is lit at night. In Newport, a locally led master lighting plan was initiated to review the architectural and public lighting, which had posed a threat to the dark sky resource. After community consultation with dedicated lighting designers, a pilot project was unveiled on a cold, crisp November evening at the close of the Mayo Dark Sky Festival 2022. The local café provided hot chocolate as the crowd gathered in the town centre and lanterns, hand-made by local school children, were distributed to all participants. A lantern walk of approximately 300 metres led a procession to the church for a musical interlude and the launch of the dark-sky-friendly lighting scheme (Figure 14.2).

Inside the church, lighting designer Kerem Asfuroglu (see his chapter in this volume for further details of the scheme) explained the ethos behind his design,

Figure 14.2   Conor McKeown (2022) *Lantern Walk, Mayo Dark Sky Festival 2022*. Use of photo with parental consent.

including issues in relation to visual light pollution and its impacts on local flora and fauna, notably, its threat to the swifts nesting in the church façade which was previously intensely floodlit.

The community remained inside the church for the musical event, whilst Kerem and the local team worked outside in getting the demonstration ready for the crowd to exit. This focused on the discrete illumination of the front doorway, with just enough light to reveal stone carvings on the Romanesque arch. Only two days before the festival, the site was surrounded by white LED floodlights, beaming up into the sky and well beyond the tower of this historic building. Artificial light shone up into, instead of down from, the heavens all night, every night:

> For the first time, the community could see stars behind the church when looking up at its façade. Jupiter shone brightly to the south of the waxing moon, which in turn cast its own beams upon those who stood to gaze.
> (Journal extract—Georgia MacMillan)

This lighting project had significant meaning to local people and represents one example of the symbolic values placed upon artificial light in some communities. The removal of the church floodlighting and implementation of a more sensitive lighting design created through community engagement was a significant shift in how this rural community viewed and engaged with the use of artificial light drew surprisingly emotional responses, an essential ingredient in developing a sense of place (Grimshaw and Mates, 2022):

> That was the moment [On turning to look back at the doorway]
> "Oh". [silence] "That is beautiful".
> A local businessman and church goer reflected that in all the 30 years he had regularly attended this church, he had never noticed the carvings in the stone arch, and he remarked upon their beauty. It was an emotional moment for him, renewing his gaze upon his own locality in a new light.
> (Journal extract—Georgia MacMillan)

In this case, the community has influenced the design for how its local nightscape can be lit and provides an example of how the dark sky paradox explored by Escario-Sierra (2022) can be successfully balanced between lighting for safety and lighting for darkness. Such projects provide pathways to bring fresh dimensions for community participation in dark skies, resonating with diverse interest groups and indirectly enabling a platform for DST.

Where next?

Using DST as a vehicle for sustainable tourism, this chapter has explored therapeutic nightscapes, indigenous knowledge, and community participation. The points raised in relation to therapeutic nightscapes posit potentially transformative, awe-inspiring experiences for the dark sky tourist. Yet it remains to be seen as to whether a reciprocal experience is also possible for the nocturnal ecosystem, and this is a critical point as without underpinning DST with measures to tackle light pollution, the future of DST is vulnerable. It is undoubtedly an exciting time for DST as the attraction to physical spaces of darkness increases, but other factors, such as the drive to secure renewable energy resources, threaten its future as the consumption of artificial light becomes cheaper. It is therefore incumbent on those seeking to develop DST to promote transferable pro-environmental behaviours towards darkness and its place in society, nature, and the sustainable tourism economy. Having engaged with the phenomenon of natural darkness, we have considered the symbiotic dependency upon darkness as a resource and as an intrinsic cultural asset. Restoring the traditional relationships that communities once had with the night sky makes a valuable contribution to the preservation of oral histories, stories, and starlore, and empowers local custodians to develop authentic and sustainable tourism experiences. To implement this, a set of principles for developing DST could be modelled on other frameworks for responsible outdoor recreation, such as the seven principles of Leave No Trace, which provides an outdoor ethics programme to minimise any impact on the environment. However, the key points of such a framework for DST must also draw from the collective and collaborative knowledge of community stewards, practitioners, and agencies in developing innovative strands.

In its opening lines, this chapter presents us with an image of home to show how far humankind has pushed the limits of travel and tourism. Through reconsidering what it is that DST can provide, we can promote an achievable form of sustainable tourism and identify with the most basic of natural phenomena: a dark night sky, wherein we can foster our own experience of nocturnal space on this pale blue dot. After all, there is no place like home.

References

Auala, S., Dalgleish, H., Shiweda, N. and Backes, M. (2023) 'Can dark sky tourism benefit Namibia?', *EASST Review*, 42(1). Available at: https://easst.net/easst-review/easst-review-volume-421-july-2023/can-dark-sky-tourism-benefit-indigenous-communities-in-namibia/.

Azevedo, A. (2018) 'Lighthouse tourism: Is there a "dark" side?' *International Journal of Tourism Cities*, 4, pp. 54–67.

Bell, S., Foley, R., Houghton, F., Maddrell, A. and Williams, A. (2018) 'From therapeutic landscapes to healthy spaces, places and practices: A scoping review', *Social Science & Medicine*, 196, pp. 123–130.

Binneman, A. and Davis, C. (2020) 'Star stories: Using indigenous knowledge for stakeholder engagement', *Communitas*, 25, pp. 1–17.

Blair, A. (2016) *Sark in the dark. Wellbeing and community on the dark sky Island of Sark.* Lampeter: Sophia Centre Press.

Blair, A. (2018) 'An exploration of the role that the night sky plays in the lives of the dark sky Island community of Sark', *Journal of Skyscape Archaeology*, 3(2), pp. 236–252.

Blundell, E., Schaffer, V. and Moyle, B. (2020) 'Dark sky tourism and the sustainability of regional tourism destination', *Tourism Recreation Research*, 45(4), pp. 549–556.

Brewer-Smyth, K. (2022) 'The healing power of nature on the brain: Healing in Green spaces and blue spaces', *Adverse Childhood Experiences*, pp. 413–427.

Boluk, K., Cavaliere, C. and Higgins-Desbiolles, F. (2019) 'A critical framework for interrogating the United Nations sustainable development goals 2030 agenda in tourism', *Journal of Sustainable Tourism*, 27(7), pp. 847–864.

Daddi, T., Bellini, N., Khan, O. and Khanna, M. (2020) *Literature review on sustainable tourism*. Spain: Consulta Europa Projects and Innovation S.L/CE.

Dalgleish, H. and Bjelajac, D. (2022) 'Dark sky tourism', in Buhalis, D. (ed.) *Encyclopedia of tourism management and marketing*. Cheltenham: Edward Elgar Publishing.

Dalgleish, H., Mengistie, G., Backes, M., Cotter, G. and Kasai, E. (2021) 'Dark sky tourism and sustainable development in Namibia', *Proceedings of the International Astronomical Union*, 15(S367), pp. 360–362.

Davies, T., Bennie, J., Inger, R. and Gaston, K. (2013) 'Artificial light alters natural regimes of night-time sky brightness', *Scientific Reports*, 3(1), p. 1722.

Derrien, M. and Stokowski, P. (2020) 'Discursive constructions of night sky experiences: Imagination and imaginaries in national park visitor narratives', *Annals of Tourism Research*, 85, p. 103038.

Dowd, M. and Hensey, R. (2016) *The archaeology of darkness*. Oxford and Philadelphia, PA: Oxbow Books.

Dunn, N. and Edensor, T. (2020) *Rethinking darkness: Cultures, histories, practices*. London: Routledge.

Edensor, T. (2013) 'Reconnecting with darkness: Gloomy landscapes, lightless places', *Social & Cultural Geography*, 14(4), pp. 446–465.

Eklöf, J. (2022) *The darkness manifesto: How light pollution threatens the ancient rhythms of life*. London: Random House.

Escario-Sierra, F., Álvarez-Alonso, C., Moseñe-Fierro, J. and Sanagustín-Fons, V. (2022) 'Sustainable tourism, social and institutional innovation—the paradox of dark sky in astrotourism', *Sustainability*, 14(11), p. 6419.

Eslami, S., Khalifah, Z., Mardani, A., Streimikiene, D. and Han, H. (2019) 'Community attachment, tourism impacts, quality of life and residents' support for sustainable tourism development', *Journal of Travel & Tourism Marketing*, 36(9), pp. 1061–1079.

Fernández-Llamazares, Á. and Cabeza, M. (2018) 'Rediscovering the potential of indigenous storytelling for conservation practice', *Conservation Letters*, 11, p. e12398. https://doi.org/10.1111/conl.12398.

Fitzgerald, P. and Lambkin, B. (2008) *Migration in Irish history 1607–2007*. London: Palgrave Macmillan.

Foley, R. and Garrido-Cumbrera, M. (2021) 'Why green and blue spaces matter more than ever', in Andrews, G. J., Crooks, V. A., Pearce, J. R. and Messina, J. P. (eds.) *COVID-19 and similar futures: Pandemic geographies*. Cham: Springer International Publishing, pp. 281–289.

Foster, M. (2017) 'Travel trend: Astro-tourism', *Green Suitcase Travel*, 2 September. Available at: https://greensuitcasetravel.com/2017/09/tourism-trend-astro-tourism/ (Accessed: 13 December 2022).

Frost, J. and Frost, W. (2021) 'Exploring prosocial and environmental motivations of frontier tourists: Implications for sustainable space tourism', *Journal of Sustainable Tourism*, pp. 2254–2270. https//doi.org/10.1080/09669582.2021.1897131.

Gajdošíková, Z. (2023) 'Sustainable tourism and degrowth: Searching for a path to societal well-being', in *Tourism, travel, and hospitality in a smart and sustainable world*. Cham. Springer International Publishing, pp. 159–173.

Grimshaw, L. and Mates, L. (2022) '"It's part of our community, where we live": Urban heritage and children's sense of place', *Urban Studies*, 59, pp. 1334–1352.

Hamacher, D., De Napoli, K. and Mott, B. (2020) 'Whitening the sky: Light pollution as a form of cultural genocide', *Journal of Dark Skies Studies*. https://doi.org/10.48550/arXiv.2001.11527.

Hensey, R. (2021) *Rediscovering the winter solstice alignment at Newgrange, Ireland.* Oxford: Oxford University Press.

Jacobs, L., Du Preez, E. A. and Fairer-Wessels, F. (2020) 'To wish upon a star: Exploring Astro Tourism as vehicle for sustainable rural development', *Development Southern Africa*, 37(1), pp. 87–104.

Kelly, C. (2018) '"I Need the Sea and the Sea Needs Me": Symbiotic coastal policy narratives for human wellbeing and sustainability in the UK', *Marine Policy*, 97, pp. 223–231.

Khetrapal, N. and Bhatia, D. (2022) 'Our brightly-lit future: Exploring the potential for astrotourism in Khajuraho (India)', *The Canadian Geographer*, 66(3), pp. 621–627.

Marr, N. (2020) 'Contact zones: The Galloway forest dark sky park as creative milieu', in Dunn, N. and Edensor, T. (eds.) *Rethinking darkness: Cultures, histories, practices.* London: Routledge, pp. 138–149.

Mayo Dark Sky Park. (2022) *Mayo dark sky festival* [Online]. Dublin, Ireland: National Parks and Wildlife Service. Available at: www.mayodarkskypark.ie/what-s-on/mayo-dark-sky-festival (Accessed: 2 June 2023).

Pereira, J., Archibald, K. and Dalgleish, H. (2022) 'Access to the night sky as a human right', 56th International Astronomical Youth Camp Report Book.

Rogers, J. and Sovick, J. (2001) 'The ultimate cultural resource?' *The George Wright Forum*, 18(4), pp. 25–29.

Sagan, C. (1994) *Pale blue dot: A vision of the human future in space.* New York, NY: Random House Publishing Group.

United Nations. (2018) *Cultural rights and the protection of cultural heritage.* A/HRC/RES/37/17. Available at: https://digitallibrary.un.org/record/1486785?ln=en (Accessed: 16 March 2023).

United Nations. (2022) *The human right to a clean, healthy and sustainable environment.* A/RES/76/300. Available at: https://digitallibrary.un.org/record/3983329?ln=en (Accessed: 16 March 2023).

UNWTO. (2013) *EU guidebook on sustainable tourism for development* [Online]. Madrid: United Nations World Tourism Organization (Accessed: 2 June 2023).

Weaver, D. (2011) 'Celestial ecotourism: New horizons in nature-based tourism', *Journal of Ecotourism*, 10(1), pp. 38–45.

15 Nocturnal (dark) anthropology

Spotlight on an ancient Indian civilisation

Neha Khetrapal

Introduction

Artificial light at night (ALAN) and night-time illumination has had a long association with security and more recently—with beautification of cities after sunset and before sunrise. Beyond safety and decoration, the time of the night is reserved for sleep and rest, while the time of the day is meant for work and production. For centuries, nights without lights have been utilised by people for activities that could not be conducted in the broad daylight, such as burgling, hunting, and praying. The antagonistic relationship that the night and the day shared necessitated the domestication of night with the help of light. With the passage of time and with fancy illumination, societies have lost touch with the night and the natural sources of light at night, which includes the starlight and the moonlight.

However, modern times have witnessed the resurrection of night as a legitimate area of investigation for ethnographers, geographers, and anthropologists. Night studies have emerged as a theme in anthropology (Galinier *et al.*, 2010; Schnepel and Ben-Ari, 2005), which had previously been dominated by a day-centred tradition. These are exciting times for anthropologists and associated scholars with the shift in investigative focus entailing the adoption of new tools and paradigms for studying spatiotemporal nocturnal dimensions. In exploring the nature of aesthetics and imaginaries stirred by the nocturnal hours, this chapter examines the rituals, beliefs, and imaginaries that are broadly situated within an Indian context, supplemented by comparative examples from other cultures.

Imagination and imaginaries

Imagination is both an individual and a social process that contributes to a sense of cultural reality. In several ways this includes creating images or representations of objects that do not exist or of objects that do exist but are not immediately perceptible (such as celestial objects dotting the skies) (Ricoeur, 1994). Imaginaries may be accessed through verbal expressions, the artefacts that people produce, or graphic, pictorial, and written products (Salazar, 2020). Deeply intertwined with cultural and social contexts, imaginaries play an important role because they bind people together as parts of larger groups (see Dobbernack, 2014). In this context,

DOI: 10.4324/9781003408444-21
This chapter has been made available under a CC-BY-NC-ND 4.0 license

the chapter underscores how imaginaries are at the core of nocturnal anthropology. When examined historically and by focusing on the night sky, imaginaries help researchers to explore how ancient societies described the night sky and celestial phenomena. Several other questions are raised. What are the most meaningful aspects about night skies? Would societal wellbeing suffer if people lack access to these meaningful aspects? How would night imaginaries be configured if people cease to engage with night skies? Do cultural variations influence or shape cultural imaginaries or are there widespread consistencies across space and time?

This chapter is divided into four parts. The first part elucidates ancient celestial observations accessed from the historical records of various cultures, with a special emphasis on the Indian context. The second part illustrates how night imaginaries are transmitted across generations through material forms (seals, carvings, and architecture), with instances from the Indian subcontinent highlighted. The third section focuses on how the differential, changing significance of the night skies might lead to the loss of both associated imaginaries and material forms. The last section underscores the contemporary means of re-establishing or probing people's changed relations with the night skies and draws contemporary implications for preservation and economic activities.

Ancient celestial imaginaries from historical records

The skies, especially dark skies, have served as a considerable source of inspiration for human ancestral groups and their imaginaries. For centuries, people have keenly examined and observed the night skies and have both externalised the results of their examinations as built architectural forms and have internalised these as religious beliefs or theological insights. This illustrates the predisposition for people to find patterns or perceive meaning in the absence of externally supplied patterns—a tendency labelled as pareidolia by psychologists (Liu *et al.*, 2014). When it comes to constellations or random groups of stars in the night skies, pareidolic tendencies have motivated people to attach diverse meanings to celestial bodies. For instance, the Pleiades, a cluster of seven blue stars in the constellation of Taurus visible from most parts of our planet, have been a potent subject for religious beliefs. The Pleiades are also the subject of an ancient Greek mythical story about the seven daughters of the Titan Atlas (Norris, 2016). The Seven Sisters were pursued by Orion, another star constellation visible in the Northern Hemisphere's night sky from November to February and represented as a hunter. The sisters were turned into stars to protect them from Orion.

The Greeks are not alone in interpreting Orion in this way. Several Aboriginal clans in Australia also construe Orion as a hunter, or as a young man, and also associate the Pleiades with a group of mythical women or girls who are chased by Orion (see Johnson, 2011). Several other cultures from different geographical regions have created similar stories wherein Orion chases the stars of Pleiades (Gibbon, 1972). These common narrative themes could be attributed to the movement of celestial objects that follow the path of Pleiades across the night sky (Johnson, 2011). On the other hand, psychologists may be inclined to correlate these themes with attempts at humanising both the celestial bodies and cosmic phenomena (Haynes, 1990)—undergirded by "collective societal pareidolia".

In contemporary times, literature reports that one of the seven stars—one of the Seven Sisters, Pleione—is barely visible to an unassisted human eye due to the glare from a nearby star. This was not the case 100,000 years ago when human ancestors were emerging from the African continent and migrated across Europe and Asia before eventually moving to Australia (Norris and Norris, 2021). To our ancestors, all seven stars were visible in the night sky. In line with the hypothesis on this subsequent invisibility, several mythical stories about the "lost Pleiad' have been narrated in Asia, Africa, Australia, and Europe, each offering explanations that apparently seek to reintegrate the invisible seventh star with the constellation's visible stars. For instance, in Hindu mythology the Pleiades are known as the *Krittikas*. The Krittikas, or the Seven Sisters, are married to seven sages, the latter the personification of the Big Dipper asterism, or *Saptarshi*, that is a part of the Ursa Major constellation. One day, six of the Seven Sisters fell for the charms of the fire god, Agni. Consequently, the husbands charged their wives with unfaithfulness and banished them. They eventually became the stars of Krittikas, with only one sister, Arundhati, remaining in her original position (Vahia, 2011).

This brief review shows that cultures seem to have recognised celestial groupings and used these groupings to construct imaginaries that reveal how the night skies have provided a canvas for people to weave celestial themes into their stories consistent with their predisposition for seeking patterns. Moreover, these imaginaries shaped people's terrestrial lives, and the following section illustrates how imaginaries were channelled into material forms, including coins, figurines, temples, and seals.

Celestial imaginaries and material forms

Prior to drawing connections between imaginaries and material forms, I introduce ancient religious beliefs from the Rig Vedic period (1500 to 1000 BC) as well as the post-Vedic period (1000 to 600 BC) (Dokras, n.d.). The Bronze Age civilisation— the Indus Valley civilisation—that flourished during the pre-Vedic period in the Indian subcontinent is also known as the Harappan civilisation after the city of Harappa. The civilisation thrived near the Indus River during the fourth/third millennium BCE in what is now the eastern parts of Pakistan and north-western India (Brahmavaivarta and Puranashave, 2009; Gonlin and Nowell, 2017). The earliest phase dates back to 3500 BCE while the mature phase is approximately pinned from 2700 BCE to 1900 BCE (Marsh, 2017).

The religious and economic life of the people from the Indus Valley was heavily influenced by celestial bodies. These ancient people were keen observers of lunar cycles and the movement of star constellations across the night sky. The observed cyclical patterns were integrated into mythic traditions as they gradually adopted agricultural practices and moved away from hunting or food gathering activities. The fertile land of the valley may have played a key role in precipitating this occupational shift, yet the sustenance of agricultural activities required careful time measurement to determine appropriate periods or seasons for planting and harvesting. As such, the celestial cyclical patterns may have formed an important basis for measuring time; indeed, the discovery of artefacts with astronomical notation supports this line of explanation, with discoveries resembling finds from other ancient agricultural societies, such as those identified in settlements surrounding the Nile (de Heinzelin, 1962). A common Indian form of marking depicted a lunar phase that consisted of a cycle of waxing moon (13 days), full moon (3 days), a waning period of 11 days, and an invisible moon period for 1–2 days. Such designs were imprinted on ancient pieces of pottery and later, on Nicobar sticks discovered from Nicobar Island (Rao, 2005).

The impact of the lunar phases and other cyclical patterns associated with star constellations extended beyond the agricultural domain, to the celestial imprints on the religious life of people, notably in deploying iconographic features of material forms upon seals, figurines, and temples. Since the domain is broad, the analysis of celestial imprints is restricted to two themes. First, the focus is on the ancient worship of *Śakti*, or Mother Goddess, that marks a seminal association between fertility, the menstrual cycle, and the observations of lunar phases. Second, an explanation is adopted that seeks to showcase how the later introduction of the sun god and other associated social changes impacted the later-occurring incarnation of Mother Goddess.

The ancient people of the Indus Valley, like their counterparts in Europe and Australia, monitored constellations and the moon's monthly cycle. Initially, these observations may have been undertaken to support foraging or hunting, but with the discovery of fertile land, the relationship between celestial movements and the fertility of nature became marked. Two core elements centred upon the cyclical features of natural phenomena that have a beginning and an end and a corresponding relation between the moon and the menstrual cycle of women (North, 2008). Terrestrial life thus became intimately linked to celestial phenomena, with the observations of natural phenomena and their celestial themes and cyclical patterns infusing social and religious imaginaries and shaping the material forms that the ancient people left behind. The following two sub-sections discuss a few prominent seals and figurines from the era.

Lunar divinities and religious life

The pre-Vedic period was characterised by the predominance of female deities. The worship of Śakti, or the feminine creative force, is one of the longest standing traditions, known from prehistoric times. The term *Śakti* embodies power or dynamic energy that is responsible for creation like mothers who give birth to a new life. Like the natural and celestial cyclical patterns, Śakti also maintains and destroys. The cycle of creation, maintenance, and destruction continues indefinitely. The seminal connection to the lunar phase is laid out by Roy (1976), who states that the half-crescent moon was the symbol of mother goddesses from several cultures. A comparable example is the Greek moon goddess, Selene, who has been depicted as a woman with a crescent moon placed on her head.

Several texts mention that the notion of feminine power that undergirds the goddess Śakti subsumes both erotic and maternal aspects (Wangu, 2003), as illustrated by the excavated Indus-clay female figurines and stamp seals—bearing images that highlight female reproductive organs (Sullivan, 1964). The latter images possibly visualise fertility rather than sensuality, although their interpretation is open to debate.

The erotic and the regenerative potential of Śakti may also accrue from its association with the cult of *Śiva*—specifically the symbolism of Śiva as a phallic form—even though De (2021) argues that Śaivism should not be simply considered a sexual cult. Indeed, other forms of Śaivism have also been prevalent, such as the tantric form of Śaivism that thrived in the second half of the first millennium (see Hatley, 2012) and attributes a terrifying character to Śakti.

At any rate, Śakti stimulates Śiva, who embodies passive energy in the form of consciousness. For the Indus Valley people, Śiva was also Pasupati or Trimukha (Singh, 2011) and was worshipped as a phallic deity along with Mother Goddess (Ferreira, 2019). Interestingly, there appears to be an intrinsic relationship between phallic gods and female deities in various other cultures across history, including the classical civilisations of Egypt, Greece, and Rome (Bachofen, 1967).

The major support for the worship of Śiva comes from Seal 420 that was discovered at the Indus Valley civilisation site by archaeologists John Marshall and Ernest Mackay in the late 1920s. The central and largest element of the seal is a human-like figure, wearing a headdress with horns and seated on a platform in a yogic position (Figure 15.1). Two wild animals are etched on the either side of the figure—an elephant and a tiger on one side and a water buffalo and a rhinoceros on the other. Besides the animals, there is also an etched fish. The central figure is carved with a quasi-phallus between its legs. John Marshall describes the figure as a prototype of Lord Śiva. Marshall further explains that the four animals highlight Śiva's character as the Lord of the Beasts (Srinivasan, 1975). This interpretation is also in line with the other activities carried out by the inhabitants of Indus Valley, for animals were domesticated to support agricultural production.

Thoughtful objections have periodically been expressed to counter Marshall's interpretation of the central figure that appears on Seal 420 (see Saletore, 1939; Sullivan, 1964; Ferreira, 2019, for other explanations). Despite these objections, two key features link it to lunar veneration and these aspects are more important for the present argument. First, the horned motif also appears on other pieces of material arts belonging to the Indus Valley (Ferreira, 2019). Second, the fish symbol—as it appears on this seal—is also visible on other seals and in the Indus script (Mahadevan, 2011).

The horned motif is intricately linked to lunar veneration, fertility, and harvest and may have been older than the Indus tradition. Pre-Harappan pieces of pottery from the site of Kot Dijian also depict a horned head. Both horns are marked with white paint marks corresponding to the lunar count, with the size of marks decreasing from the bottom to the tip to signify the phases of the moon. Beyond the Indus Valley, evidence from other ancient societies supports this proposed lunar link. A prehistoric French Venus of Laussel is etched with amplified bodily parts.

Figure 15.1   Seal 420 at the National Museum, New Delhi. Creative Commons.

She holds a crescent-moon-shaped bison horn in one hand that bears 13 vertical marks that possibly signify 13 lunations in a year, or the rhythm of ovulation (Marler, 2003). And from ancient Mesopotamia is a bull, a specifically lunar animal with horns representing the crescent moon (Ornan, 2001). Across cultures, the widespread use of bulls for ploughing may constitute the underlying factor that precipitated the association between the moon and the bull. At any rate, the commonalities across ancient cultures shows that people's attempt to make sense of their celestial surroundings and their association of celestial phenomena with their common agrarian needs generated similar imaginaries across different geographical areas. Consistently, material forms from different cultures bear similarities as well.

The examples from the Indus Valley show that as with other cultures, celestial objects from the dark skies left deep imprints on religious and social life. As far as the specific relationship between Śakti and Śiva is concerned, Śakti played a more prominent role while the Proto-Śiva may have been her consort (Ferreira, 2019). Ancient lunar worship is also visible in contemporary depictions of Śiva in which he is portrayed with a crescent moon placed like a crown on his forehead. Alternatively, it may be that the crescent moon replaced the horned headgear after the pre-Vedic period, as the worship of Śiva became more prominent (Rao, 2005).

Secondly, in several older linguistic notations from India, the fish symbol signifies a star (Heras, 1953). Close scrutiny of the Indus text reveals a combination of 6 fishes that have been interpreted as associated with the six stars of Pleiades

(Mahadevan, 2011) that as previously discussed, played a significant role in ancient Indian mythology.

The Bull Cult and the taurus constellation

The Pleiades, due to their easy visibility and with an annual setting and rising pattern after a complete orbit of earth, earned a special place amongst the Indus Valley inhabitants. As the previous text shows, the significance of Pleiades can be gauged through its significance for marking agricultural activities that accrue from the star cluster's positions in the dark sky, marking the beginning of autumn and spring or seasons for ploughing and seeding (Ceci, 1978). While direct evidence is not available to link cattle breeding with the cyclical visibility of the Pleiades as it may be available for other ancient nomadic cattle groups (Sparavigna, 2008), the discovery of seals with animal motifs paired with the use of beasts for support-ing agricultural production hints at links between the star cluster and cattle for the Indus Valley people.

A further line of distal evidence is available in the form of legendary stories that associate the goddess Śakti, the Seven Sisters, and bull veneration, mirroring the spatial proximity between the Pleiades and the constellation of Taurus, with both generally observed in the Northern Hemisphere's winter months (Ceci, 1978). An impressive feature of Taurus is its V-shaped pattern that resembles a bull's head and is called Hyades. Another bright, reddish star, Aldebaran, lies very close to Hyades, making it easy to perceive Aldebaran as the bull's red eye. The celestial bull impressions are captured both in material forms that the Indus Valley people produced and the mythical tales that they left behind. Examination of Indus Val-ley figurines (see Figure 15.2), pieces of pottery and seals reveal that the bull was not only a key figure (Sharma, 2018) but has also been closely associated with the V-shaped pattern of Hyades. The latter claim finds support in the depiction of bull horns that are commonly represented as a V (Hiltebeitel, 1978), with the most majestic motif—the zebu bull—with a hump above its shoulders.

Beyond simple pictorial depiction, the bull mythology becomes quite complex. On one hand, some argue that the crescent moon worn by Śiva is intricately linked with bovine symbolism rather than the moon or any other celestial asterism (Hilte-beitel, 1978), an argument that foregrounds the social importance of the bull as a domesticated animal and as a mode of transportation (Sharma, 2018). On the other hand, others argue that Śiva's bull epithet emanates from the traits and features of his predecessor, *Rudra*. The storm deity of that time, Rudra, is popularly labelled as the wild bull of heaven or simply the great bull (Long, 1971). The resonance with the ancient bull-Śiva pairing is the close spatial depiction of a white bull and *Śiva* in several present-day temples across India.

Other cultures around the world have also had a place for legendary bull stories, which have often been translated into cave paintings and figurines, as for exam-ple, with the paintings of the outlines of a bison in the ancient Palaeolithic cave discovered in south-western France. In addition, the fertility cult and the concept of Mother Goddess has been closely associated with bull symbolism within other

Figure 15.2 Animal figurines at the National Museum, New Delhi. Photograph by Neha Khetrapal.

Neolithic cultures. Striking images include those of a Mother Goddess with a bull-headed son or a child (Ganesh, 1990; Sharma, 2018). Several historians and theorists contend that the bull-headed son may represent the Hyades or the constellation of Taurus (Sołtysiak, 2001).

The bull-son finds parallels in the Greek myths of Mother Goddess or Mother Earth too (Dietrich, 1967) that focus on the story of a divine child, bearing close resemblance to the bull, who is born and dies annually, much like the cycle of the growth and decay of vegetation. These tales share features with the cycle of creation, maintenance and destruction that are often attributed to Śakti. Although there are interesting similarities, a direct comparison between the divine child and the child of Śakti remains to be drawn. The son of Śakti, not introduced thus far, is known by various names—Skanda, Kārttikeya, Murugan, and Kumāra. Evidence drawn from both textual and non-textual sources—seals and coins—traces the cult of Skanda to North India. According to the famous Indus Valley hypothesis (Agrawala, 1967; Mann, 2007), the term *Kārttikeya* alludes to the possibility that Skanda was the foster son of Krittikas, who may have evolved from Seven Sisters to seven divine mothers by this time. Although there are varied iconic representations of Skanda, the one with six heads offers links between him and his six mothers who are visible in the night sky to an unaided eye. Unearthed Indus seals depicting Skanda with the seven divine mothers (including the lost Pleiad) corroborate this line of argument (Aravamuthan, 1948). Within this framework of Skanda's origin,

Figure 15.3 The Dancing Girl statuette at the National Museum, New Delhi. Photograph by Jen Calliope. Creative Commons.

a rare astrological event that occurred during the third millennium BCE may have been responsible for Skanda's birth (Volchok, 1970).

From the period between the first and the fourth century CE (Dasgupta, 1974), Skanda is also portrayed as a warrior, the brave son of Śiva and his consort, Parvati. As a warrior, Skanda is associated with Orion, who slays the buffalo-demon (Mahishasura) with his spear in the view of Śiva, as portrayed in the Harappan tablet H95–2486. In this claim to Skanda's parentage, Śiva's dominance is asserted over the others, including his consort Parvati, the embodiment of Śakti. According to a recent controversial claim (Chopra, 2016), the famous Dancing Girl statuette from the Indus Valley represents Parvati (see Figure 15.3).

At the same time, it is hard to ignore arguments that there is no connection between Skanda and Śiva (Agrawala, 1967). To reconcile these opposing lines of arguments, others have claimed that Skanda may have been absorbed into the cult of Śiva by the seventh century (Chatterjee, 1970). The absorption of Skanda is paralleled by the modified iconography that emerged after the third millennium BCE, where Skanda is more frequently paired with Śiva. In the Southern Indian Tamilian iconography, Skanda's mother, Parvati, possesses a green complexion, exemplifying her continuity with nature, Mother Goddess, and the cycle of rejuvenation and destruction on Earth—as signalled by the rising and setting of Pleiades.

Influx of Aryans and modified ancient imaginaries

The Indus Valley people came in contact with several Indo-European groups between 1750 and 400 BCE (Wangu, 2003), who entered India via difficult routes over mountain ranges and failed to bring a comparable number of females with them (Sarkar, 1987). These nomadic groups, also called Aryans, appeared to be in conflict with the people of Indus Valley. This conflict eventually gave way to cooperation (Wangu, 2003). Nevertheless, Aryans who are believed to be worshippers of nature, introduced new gods into the social fabric of the region. One of the most important forms of worship was associated with the sun god (Jing, 2018), with a wheel or a golden circular plate as a key symbol (Hopkins, 2005). The wheel appears prominently in various sun temples later built across India, exemplifying how Aryan imaginaries, beliefs, and rituals were translated into material forms. However, the dominance of the Aryan groups also meant that the imaginaries that originated from the Indus Valley and their associated material forms were assigned a differential level of importance or relegated to the periphery. Consequently, visual images of female divinities seem to be absent during the period under discussion (Mukherjee, 1983). With the further passage of time, male divinities became centralised. Put differently, the dark skies that structured the religious and social landscape of ancient Indian society were subsumed under bright skies in much the same manner as the struggle between the Neanderthal religion, characterised by moon worship, and the sun religions, which later became prominent (Laoupi, 2006).

The transformation was brought about by the influx of Aryans, whose deities, representing the gendered fabric of their clans, were predominantly male (Sahgal, 2018). As a result, the Aryan patriarchal stratum superseded the matriarchal stratum of the Indus Valley (Fane, 1975). Thereafter, mother goddesses were marked by their peripheral importance (Dasgupta and Ashrama, 2004). Evidence for the dwindling importance of female deities accrues from the religious-philosophical texts of Hinduism, the Upanishads, that pay scant attention to female deities in the post-Vedic period, except for their roles as saviours to men or as wives of male deities (Amazzone, 2011). The major male deity to assume prominence was Śiva, whose transformation from proto-Śiva to Śiva exemplifies the transformed gendered-structuring of the era (see Lodrick, 2005, for other explanations associated with the prominence of Śiva). With the rising popularity of Śiva, Krittikas—identified with the stars of Pleiades—were incarnated as well. Identified as foster

Figure 15.4 Sapta-Matrikas from Pratihara, Uttar Pradesh (dated, ninth century CE) at the National Museum, New Delhi. Photograph by Rohini. Creative Commons.

mothers to Skanda, their cluster was elaborated as the Matrikas in Puranic litera-ture (Goswami, Gupta and Jha, 2005)—whose composition dates from the fourth century BCE to 1000 CE. In the form of Sapta-Matrikas, or Seven Divine Mothers (see Figure 15.4), these female deities were also considered the goddesses of the battlefield—as created by Śiva—and who fought demons alongside him.

Although space restricts the elaboration of these transformation stories, a few important summary points are noteworthy. First, the influx of Aryans transformed the matriarchal structure of the Indus Valley. Second, the links between the venera-tion of night-time celestial bodies and female divinities dwindled with time. Third, the prominence of male divinities implied that female divinities played roles that accorded with their positions as consorts of male deities. Fourth, the introduction of sun worship inspired new material forms (sun temples) that become more popu-lar than the material forms and architecture associated with female divinities.

Bright skies: sun worship

Despite this struggle and subjugation, the worship of Śakti and the reverence for Matrikas were handed down to the next generations beyond the Puranic period, albeit in a modified form and paired with the respect that was offered to the bright skies (see Das, 1977, for the spread of the Śakti cult). For instance, the Indus Val-ley seals, depicting female sexuality and fertility, are comparable to the images that belong to the early Gupta period (Banerjea, 1954) from the early fourth century CE to the late sixth century CE. A famous example is the image of Lajja Gauri, a female deity that bears a lotus head and exposed genitalia (Sankalia, 1960). Other discoveries of Indian plaques dating back to the fifth century CE links Lajja Gauri with the white bull icon of Śiva. These discoveries also link the worship of Lajja Gauri to the fertility cults from the Harappan times (Mishra, 2004).

Simultaneously, the sun god had achieved the status of the chief deity by the Gupta period, and several sun temples were built (Sarkar, 2011). Beyond the Gupta period, there is evidence for Matrika sculptures from the periods of the Pallavas, Cholas, and Pandyas (7th to 13th century). For these sculptures, their clothing is minimal as well (Goswami, Gupta and Jha, 2005). It is important to emphasise that although attempts to date several social and religious transitions have been carefully considered for this chapter, there are several transitions that either happened simultaneously or the historical dates—as mentioned—have been reported as estimates in previously published literature.

Revival of dark skies (contemporary)

Before the intrusion of modern ALAN and night-time illumination, ancient dark skies played a significant role in the development of religious practices, timekeeping mechanisms, architecture, and astronomy around the world. The current overview of literature from ancient Indian civilisation is only a small aspect of this global phenomenon. At the same time, astral timekeeping strategies, astral festivals, and material forms inspired by lunar traditions are other astral facets that have considerably declined over time. A large chunk of literature that tracks the decline of astral traditions and material forms is not easily accessible or has not been published, at least within the Hindu tradition. However, comparative examples from the West serve as an important basis for concluding that the relevance of night skies for Hindus and Indians has dwindled considerably. For instance, data available from Europe shows that festivals linked to star constellations are practically non-existent today (for instance, as with the festival of Panathenaia, celebrated in ancient Athens, Boutsikas, 2011). It is probable to propose that Indian night-time traditions have lost their appeal in the face of the increased reverence for the sun god and restricted access to dark skies.

More positively, the contemporary world has witnessed a revival of interest in dark skies, as testified by the recent rise of dark skies locations in Scotland (Edensor, 2013) and Arizona. Furthermore, tourism has also increased to Iceland, Canada, and Norway, where natural darkness is an essential ingredient for experiencing *aurora borealis* (Edensor, 2011). The increase in tourism and natural dark spots is encouraging, but it still remains to be seen whether this contemporary interest can compare to the widespread sanctity of ancient darkness.

Recently, Khetrapal and Bhatia (2022) argued for combining night-time tourism with cultural heritage by citing the Khajuraho temples of Central India as a means of arresting the loss of heritage value attached to architecture inspired by ancient mythical beliefs (see Figure 15.5). They further argue that the success of this new endeavour is dependent upon the minimisation of artificial illumination and the preservation of natural starlit reserves. In the long run, efforts to combine night-time tourism with cultural heritage might provide a unique opportunity to foster engagement with the nocturnal environment and rediscover the cultural or religious meanings associated with starlight and moonlight.

Figure 15.5 Temple at Khajuraho, Madhya Pradesh, India. Photograph by Paul Mannix. Creative Commons.

For the future, it remains to be seen how nocturnal imaginaries that integrate cultural practices, traditions, material forms, and architecture could potentially serve as a driving force for making progress in nocturnal, dark anthropology. In the interim, it appears as if the contemporary dichotomy between day and night may become more blurred and the metaphoric association of night with suffering and danger may diminish as endeavours within the domain of nocturnal anthropology begin to take shape.

References

Agrawala, P. K. (1967) 'Skanda-Karttikeya (A study in the origin and development)'. *Monographs of the department of ancient Indian history, culture and archaeology*. Varanasi, India: Banaras Hindu University.

Amazzone, L. (2011) 'Durga: Invincible goddess of South Asia', in Monaghan, P. (ed.) *Goddesses in world culture*. Westport, CT: Praeger, pp. 71–84.

Aravamuthan, T. G. (1948) 'More gods of Harappa', *Journal of the Bihar Research Society*, 34(3–4), pp. 31–82.

Bachofen, J. J. (1967) *Selections myth, religion, and mother right: Selected writings*. Princeton, NJ: Princeton University Press.

Banerjea, J. N. (1954) 'Some aspects of Shakti worship in Ancient India', *Prabhuddha Bharat*, 59(3), pp. 227–232.

Brahmavaivarta, D. B. and Puranashave, V. L. (2009) 'Cult of Varahi in Orissa', *Orissa Review*, pp. 80–84.

Boutsikas, E. (2011) 'Astronomical evidence for the timing of the Panathenaia', *American Journal of Archaeology*, 115(2), pp. 303–309.

Ceci, L. (1978) 'Watchers of the Pleiades: Ethnoastronomy among native cultivators in Northeastern North America', *Ethnohistory*, pp. 301–317.

Chatterjee, A. K. (1970) *The cult of Skanda-Kiirttikeya in ancient India*. Calcutta: Punthi Pustak.

Chopra, R. (2016) 'Mohenjodaro "Dancing girl" is Parvati', *The Indian Express*, 26 December. Available at: https://indianexpress.com/article/india/mohenjodaro-dancing-girl-is-parvati-claims-ichr-journal-4444981/.

Das, B. (1977) 'The development of Sakti cult in Orissa during ancient and medieval periods', *Proceedings of the Indian History Congress*, 38, pp. 130–137.

Dasgupta, K. K. (1974) *A tribal history of ancient India*. Calcutta: Nababharat Publishers.

Dasgupta, S. B. and Ashrama, A. (2004) *Evolution of mother worship in India*. Belur Math: Advaita Ashrama.

De, D. (2021) 'The aniconic cult of Lord Shiva: The mostly different and mysterious God of the Hindu pantheon', in Pathak, G. and De, D. (eds.) *Approaches into Indian history and cultural heritage*. New Delhi: Prakhar Goonj, pp. 99–115.

de Heinzelin, J. (1962) 'Ishango', *Scientific American*, 206(6), pp. 105–118.

Dietrich, B. C. (1967) 'Some light from the east on Cretan cult practice', *Historia: Zeitschrift für Alte Geschichte, (H. 4)*, pp. 385–413.

Dobbernack, J. (2014) *The politics of social cohesion in Germany, France and the United Kingdom*. London: Palgrave Macmillan.

Dokras. (n.d.) *Puranas—collection of essays on Hindu temples*. Stockholm: Indo Nordic Author's Collective.

Edensor, T. (2011) 'Aurora landscapes: The effects of light and dark', in Benediktsson, K. and Lund, K. (eds.) *Conversations with landscape*. Aldershot: Ashgate.

Edensor, T. (2013) 'Reconnecting with darkness: Gloomy landscapes, lightless places', *Social and Cultural Geography*, 14(4), pp. 446–465.

Fane, H. (1975) 'The female element in Indian culture', *Asian Folklore Studies*, 34(1) pp. 51–112.

Ferreira, M. V. (2019) *Proto-Soma and Tantra: An iconographical exploration into the possibilities of the Vedic Soma's prototype in the Indus Valley*. Long Beach, CA: California State University.

Galinier, J., Monod Becquelin, A., Bordin, G., Fontaine, L., Fourmaux, F., Roullet Ponce, J. and Zilli, I. (2010) 'Anthropology of the night: Cross-disciplinary investigations', *Current Anthropology*, 51(6), pp. 819–847.

Ganesh, K. (1990) 'Mother who is not a mother: In search of the great Indian goddess', *Economic and Political Weekly*, pp. WS58–WS64.

Gibbon, W. B. (1972) 'Asiatic parallels in North American star lore: Milky way, Pleiades, Orion', *The Journal of American Folklore*, 85(337), pp. 236–247.

Gonlin, N. and Nowell, A. (2017) *Archaeology of the night: Life after dark in the ancient world*. Denver, CO: University Press of Colorado.

Goswami, M., Gupta, I. and Jha, P. (2005) 'Sapta matrikas in Indian art and their significance in Indian sculpture and ethos: A critical study', *Anistoriton*, 9, section A051.

Hatley, S. (2012) 'From Mātṛ to Yoginī: Continuity and transformation in the South Asian cults of the mother goddesses', in Keul, I. (ed.) *Transformations and transfer of Tantra in Asia and beyond*. Berlin: Walter de Gruyter, pp. 99–129.

Haynes, R. D. (1990) 'The astronomy of the Australian Aborigines', *The Astronomy Quarterly*, 7, pp. 193–217.

Heras, H. (1953) *Studies in proto Indian Mediterranean culture.* Vol. I. Bombay: Indian Historical Research Institute.

Hiltebeitel, A. (1978) 'The Indus valley" Proto-Śiva", Reexamined through reflections on the goddess, the Buffalo, and the Symbolism of vāhanas', *Anthropos, (H.5/6)*, pp. 767–797.

Hopkins, T. J. (2005) 'Saura Hinduism', in Singh, N. K. and Mishra, A. P. (eds.) *Encyclopaedia of oriental philosophy and religion: Hinduism (volume 3).* New Delhi: Global Vision, pp. 767–768.

Jing, C. (2018) 'The mythological thought of Rigveda', *International Journal Online of Humanities*, 4(4), p. 12.

Johnson, D. (2011) 'Interpretations of the Pleiades in Australian aboriginal astronomies', *Proceedings of the International Astronomical Union*, 7(S278), pp. 291–297.

Khetrapal, N. and Bhatia, D. (2022) 'Our brightly-lit future: Exploring the potential for astrotourism in Khajuraho (India)', *The Canadian Geographer*, 66(3), pp. 621–627.

Laoupi, A. (2006) 'The Greek myth of Pleiades in the archaeology of natural disasters. Decoding, dating and environmental interpretation', *Mediterranean Archaeology and Archaeometry*, 6(2), pp. 5–22.

Liu, J., Li, J., Feng, L., Li, L., Tian, J. and Lee, K. (2014) 'Seeing Jesus in toast: Neural and behavioral correlates of face pareidolia', *Cortex*, 53, pp. 60–77.

Lodrick, D. O. (2005) 'Symbol and sustenance: Cattle in South Asian culture', *Dialectical Anthropology*, 29(1), pp. 61–84.

Long, J. B. (1971) 'Siva and Dionysos-visions of terror and bliss', *Numen*, 18(3), pp. 180–209.

Mahadevan, I. (2011) 'The Indus fish Swam in the great bath: A new solution to an old riddle', in *Bulletin of the Indus research centre, No. 2.* Chennai: Indus Research Centre, Roja Muthiah Research Library.

Mann, R. (2007) 'Skanda in epic and Puranic literature: An examination of the origins and development of a Hindu deity in North India', *Religion Compass*, 1(6), pp. 725–751.

Marler, J. (2003) 'The body of woman as sacred metaphor', in Panza, M. and Ganzerla, M. T. (eds.) *Transiti, Metamorfosi, Permanenze.* Bologna: Associazione Armonie, pp. 9–24.

Marsh, M. G. (2017) 'The Indus river valley civilization & the Vedic age of India (CA. 3000 BCE—700 BCE)', in Skjelver, D. M., Arnold, D., Broedel, H. P., Glasco, S. B., Kim, B. and Broedel, S. D. (eds.) *History of applied science & technology: An open access textbook.* Available at: https://press.rebus.community/historyoftech/part/chapter-2-the-indus-river-valley-civilization-and-the-vedic-age-of-india-ca-3000-bce-700-bce/.

Mishra, D. B. (2004) 'Headless goddess of Nuapada Orissa (A study of its antiquity and indentify)', *The Orissa Historical Research Journal*, pp. 53–64.

Mukherjee, B. N. (1983) A plea for study of art in coinage. *Numismatic Society of India* (19).

Norris, R. P. (2016) 'Dawes review 5: Australian aboriginal astronomy and navigation', *Publications of the Astronomical Society of Australia*, 33, p. e039.

Norris, R. P. and Norris, B. R. (2021) 'Why are there seven sisters?' in Boutsikas, E., McCluskey, S. C. and Steele, J. (eds.) *Advancing cultural astronomy.* Cham: Springer, pp. 223–235.

North, J. (2008) *Cosmos: An illustrated history of astronomy and cosmology.* Chicago, IL: The University of Chicago Press.

Ornan, T. (2001) 'The bull and its two masters: Moon and storm deities in relation to the bull in ancient near Eastern art', *Israel Exploration Journal*, 51, pp. 1–26.

Rao, N. K. (2005) 'Aspects of prehistoric astronomy in India', *Bulletin of the Astronomical Society of India*, 30, pp. 499–511.

Ricoeur, P. (1994) 'Imagination in discourse and in action', in Robinson, G. and Rundell, J. F. (eds.) *Rethinking imagination: Culture and creativity.* London: Routledge, pp. 87–117.

Roy, S. B. (1976) *Prehistoric lunar astronomy [19000–3100 BC]*. New Delhi: Institute of Chronology.

Sahgal, S. (2018) 'Constructing "Vedic" masculinities', *Social Scientist*, 46(9–10), pp. 23–42.

Salazar, N. B. (2020) 'On imagination and imaginaries, mobility and immobility: Seeing the forest for the trees', *Culture & Psychology*, 26(4), pp. 768–777.

Saletore, B. A. (1939) 'Identification of a Mohenjo-Daro figure', *The New Review*, 10, pp. 28–35.

Sankalia, H. D. (1960) 'The nude goddess or" shameless woman" in Western Asia, India, and South-Eastern Asia', *Artibus Asiae*, 23(2), pp. 111–123.

Sarkar, B. K. (2011) 'Temples of the sun god in early Assam', *Karatoya: NBU Journal of History*, 4, pp. 2–35.

Sarkar, P. R. (1987) 'Tantra and Indo-Aryan civilization', *Discourses on Tantra*, 1, pp. 141–175.

Schnepel, B. and Ben-Ari, E. (2005) 'Introduction: 'When darkness comes': Steps toward an anthropology of the night', *Paideuma*, pp. 153–163.

Sharma, R. (2018) 'A study of terracotta bull figurine of Harappan Sites in Rajasthan (Indian context)', *Proceedings of the Indian History Congress*, 79, pp. 744–755.

Singh, U. (2011) 'Religious life under the Chandellas', *Proceedings of the Indian History Congress*, 72, pp. 220–231.

Sparavigna, A. (2008) *The Pleiades: The celestial herd of ancient timekeepers*. Available at: http://arxiv.org/abs/0810.1592 (Accessed: 8 August 2022).

Srinivasan, D. (1975) 'The so-called proto-Śiva seal from Mohenjo-Daro: An iconological assessment', *Archives of Asian Art*, 29, pp. 47–58.

Sołtysiak, A. (2001) 'The bull of heaven in Mesopotamian sources', *Culture and Cosmos*, 5(2), pp. 3–21.

Sullivan, H. P. (1964) 'A re-examination of the religion of the Indus civilization', *History of Religions*, 4(1), pp. 115–125.

Vahia, M. (2011) *Astronomical myths in India*. Mumbai: Tata Institute of Fundamental Research.

Volchok, B. Y. (1970) 'Towards an interpretation of the proto-Indian pictures', *Journal of Tamil Studies*, 2(1), pp. 29–51.

Wangu, M. B. (2003) *Images of Indian goddesses: Myths, meanings and models*. New Delhi: Abhinav Publications.

16 Beauty won't save the starry night

Astro-tourism and the astronomical sublime

Dwayne Avery

Tourism and the starry night

Although pilgrimages to special astronomical destinations are as old as human civilisation (Brown, 2000), in recent years, astro-tourism has emerged as a popular class of ecotourism dedicated to preserving and protecting "pristine" darkness (Farajirad and Beiki, 2015; Fayos-solà *et al.*, 2014; Escario-Sierra *et al.*, 2022). With the rapid proliferation of urban light pollution, access to dark and gloomy environments free of artificial lighting is rare. Some estimate that nearly 80% of the world's population has never seen the Milky Way (Donahue, 2016), a deprivation of darkness that according to Paul Virilio (2000), represents the loss of an ancient inheritance. Indeed, while artificial lighting has benefited modern societies in various ways, the excessive illumination characterising most cities today has many questioning modernity's colonisation of the night by daytime activities (Bogard, 2014; Melbin, 1987).

Like other forms of ecotourism which use sustainable travel to bolster rural communities and foster responsible relationships with the natural world, astro-tourism promotes itself as a steward of the night. Composed of various astronomical activities, from stargazing in dark sky parks and preserves to observing rare celestial events like meteorite showers and auroral displays, celestial tourism capitalises on the beauty, cultural heritage, and scientific knowledge of the starry night to thwart the spread of malignant light pollution. Furthermore, to carry out this conservation work, many dark sky destinations take their stewardship cues from the national park conservation movements of the early 20th century. For example, some of the best places to view the Milky Way are protected dark areas found in pre-existing nature parks that have been recently certified by DarkSky, an internationally recognised authority on protecting the night sky.

In this chapter, I explore the rhetorical mythologies (Barthes, 1999) used by tourist media to construct astronomical tourist destinations. Although the rise in celestial tourism has generated a modest research literature, much of this work revolves around ontological questions about what constitutes astro-tourism and how it can bolster rural development (Weaver, 2011; Silver and Hickey, 2020; Blundell, Schaffer and Moyle, 2020, Mitchell and Gallaway, 2019). Missing from these studies, however, are important questions about the ways in which image-based

DOI: 10.4324/9781003408444-22
This chapter has been made available under a CC-BY-NC-ND 4.0 license

promotional media represent the starry night. As Albers and James (1988) show, images, in the form of photographs, postcards, and posters, are not only the primary way in which tourist destinations are created and communicated, but they are instrumental in understanding tourist experiences.

To better understand the rhetorical strategies tourist media use to construct the starry night, I analyse Tyler Nordgren's *Half the Park is After Dark*, a series of dark sky tourist posters promoting the US National Parks Service. A professional astronomer, night sky ambassador, and visual artist, Nordgren demonstrates how the battle to protect darkness depends as much on the rhetorical power of picturesque imagery as it does on scientific outreach programs. Indeed, combining a vintage aesthetic style with sublime images of the Milky Way, *Half the Park is After Dark* builds upon a century of romantic thought that envisions national parks as one of the final frontiers for experiencing uncontaminated nature. Coming alive at night, iconic landscapes, such as the Grand Canyon and Mount Rainier (Figure 16.1), are presented by Nordgren as a sublime, ancient inheritance capable of teleporting the astro-tourist back nostalgically to a time free from the blight of urban light pollution.

Although public attitudes about national parks have changed considerably since their inception in the early 20th century (Demars, 1990), Nordgren builds his mythological starry night around an antiquated romantic approach that prevents the public from perceiving the harms inflicted upon the night. While sublime darkness may offer nocturnal tourists profound and wondrous experiences with night-time, as a conservation strategy it falls short, acting more like a therapeutic service than a call to arms. Like the myth of "wilderness" (Denevan, 1992; Cronon, 1995) that has dogged national parks in the past, Nordgren treats pristine darkness as a spiritual bulwark whereby uncontaminated nature acts as the cure-all for the ills of rampant light pollution. The persistence of this conservationist outlook is easy to understand. Anyone who has stood underneath the vast expanse of the Milky Way can appreciate the magnificent and humbling feelings that come from experiencing the astronomical sublime. However, as an environmental strategy, eulogising the romantic ideals of "untouched" nature is lacking. As I show, by fetishising the starry night's ancient qualities, Nordgren's nostalgic treatment of the sublime deforms and dehistoricises the sky, treating it as an atemporal and "empty reality without depth" (Barthes, 1999). For example, instead of attuning the astro-tourist to some of the many corporate space expansion projects transforming the night, such as the booming small satellite industry, Nordgren's posters idealise the dark sky park as an "unchanging" and "authentic" antidote to the artificiality of urban modernity.

Nature parks and the myth of the pristine

National parks have been celebrated for their restorative capacities for over a century. Not only are picturesque and sublime landscapes seen as deeply enriching, capable of boosting the health and emotional life of the weary urban dweller (Ward, 2005; Patin, 2012) but throughout the early 19th century, many moral

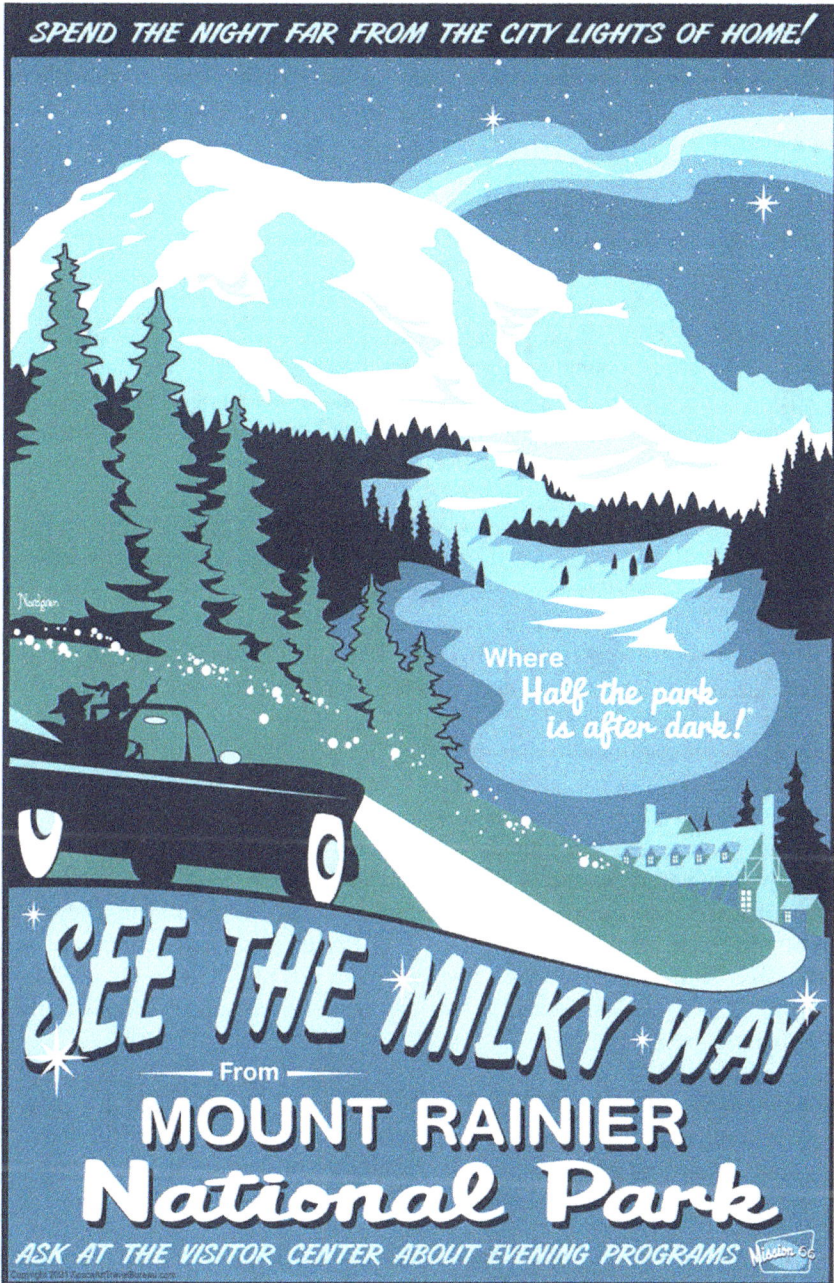

Figure 16.1   Mount Rainier National Park 1950s. Poster by Dr Tyler Nordgren.

reformers envisaged untouched wilderness as an antidote to the corrupt and morally vacant influences of urban life (Beauregard, 2002; Conn, 2014). Although secular, industrial society provided a growing list of material benefits, including the introduction of electric lighting, many felt dissatisfied with urban modernity's conception of progress, choosing instead to discover "authentic" experiences elsewhere.

To fill in the spiritual void left by the daily grind of everyday urban life, many people sought refuge in America's rugged and "untouched" western frontier. In the late 1800s, the naturalist John Muir (2020) noted,

> Thousands of nerve-shaken, overcivilized people are beginning to find out that going to the mountains is going home. Awakening from the stupefying effects of the vice of over-industry and the deadly apathy of luxury, they are trying as best they can to mix and enrich their little ongoings with those of nature, and to get rid of rust and disease

Much of the anti-modernism expressed by Muir had its roots in the Romantic movement, especially the idea that paying close attention to nature possessed many moral benefits (Demars, 1990). For example, using "new eyes" to discover manifestations of God in nature, the natural philosopher Emerson (2012, p. 39) celebrated rural landscapes for their ability to combat the moral shortcomings of industrialisation. "Our hunting for the picturesque", Emerson wrote, "is inseparable from our protest against false society".

In his classic critique of "wilderness", William Cronon (1995) describes how the environmental movement in the US proceeded along a peculiar history involving the conservation of pristine nature through a system of national nature parks. As the incessant rhythms of urban cultures came to dominate American life, Cronon notes, nature parks emerged as an ecological symbol of the purity needed to avert the spiritual collapse of modern society. To promote these newly constructed landscapes, mass tourism emerged as a powerful agent of consecration, transforming national parks, like Yellowstone, Yosemite, and Zion, into sacred places far removed from everyday urban life. Discussing the Grand Canyon's mass appeal, Mark Neumann (1999) notes,

> At the turn of this century, on the rim of its great empty chasm, a tourist world took shape that continually sought to affirm itself as a refuge from the modern world. More than anything, the Grand Canyon was a natural scene dramatizing how tourism and modernization went hand in hand.

The 19th-century idea that protected nature parks provide a moral defence against the unchecked march of urban growth continues today in tourist materials for dark sky parks and preserves. Just as the pages of previous travel guidebooks, brochures, and magazine articles encouraged the public to contemplate the sublime awe of protected landscapes, Nordgren's *Half the Park is After Dark* envisages the astro-tourist as a nocturnal adventurer whose search for pristine darkness takes them away from the drudgery of city life. For example, in the poster for Alaska's Denali National Park and Preserve, an adventurous couple stands beneath a mountainous view of the northern lights (Figure 16.2). The caption reads, "Far from city lights, dark skies save the northern lights!" The message could not be clearer: only the national park system can protect the night from the encroaching scourge of urban light pollution, a message that spreads across the entire series. Nordgren's homage to the stewardship aims of the national parks service offers a textbook case of 19th-century nature conservation, as he repeatedly takes inspiration from two of the most predominant tropes in romanticism, sublime wonder, and frontier nostalgia (Cronon, 1995).

From their inception, US nature parks have relied upon romantic feelings of sublime wonder to raise public awareness about the moral guidance of untamed wilderness. Since Nordgren's images are tourist promotions for widely recognised national parks, they already come imbued with a long history of romantic thought. However, as the series' title indicates, a different, even more astonishing version of the nature park comes alive after the sun subsides: when night falls and the Milky Way appears atop the park's grand geological formations, the onlooker may gaze upon another ancient wonder, the astronomical sublime. In most posters, Nordgren associates the astronomical sublime with the Milky Way, especially its status as a "cathedral of calm" for a mature couple, who are repeatedly found standing peacefully under a dome of ancient stars, the blue majesty of the night taking them back to the early days of human civilisation. The images elicit what the poet Henry Van Dyke (2005) called "wordless worship", a mode of quiet contemplation allowing the silence of the cosmos to facilitate an enduring emotional bond between man and nature.

While geological examples of the sublime often involve the awesome might of earthbound nature, such as the incomprehensible scale and power of mountains or raging rivers, the astronomical sublime goes further. Throughout Nordgren's posters, peering upward at the Milky Way provides a window into the meaning of the entire cosmos, a romantic sentiment conjuring Tennyson's (1987) night poet. "For the poet of the night sky . . . who looks up, not out", Tennyson wrote, "the task is to totalize something that cannot be encompassed: the infinitude of the universe itself". This "sublime of awe", as Anne Janowitz (2005) puts it, appeared in the 18th century as the predominant framework for assessing the mysteries of the night sky and celebrated the triumphs of the human mind as it comprehended the universe's vastness. In the poster for the Arches National Park, Nordgren explicitly references this sublime tradition, visualising the Milky Way as a shared resource strengthening the moral good of all human society (Figure 16.3). Alongside a magnificent scene of a lone sandstone arch pointing upward at the Milky Way, we find the caption "Windows to the Universe". In another poster, the night sky's ancient powers of unification are put on display. Alongside a pastoral image of the Milky Way reads this quote from John Muir: "We all travel the Milky Way, trees and men".

Notably absent from Nordgren's astronomical sublime however, is any reference to the potential terrors that come from looking into the "abyss" of nature. This

Figure 16.2   Denali National Park and Preserve. Poster by Dr Tyler Nordgren.

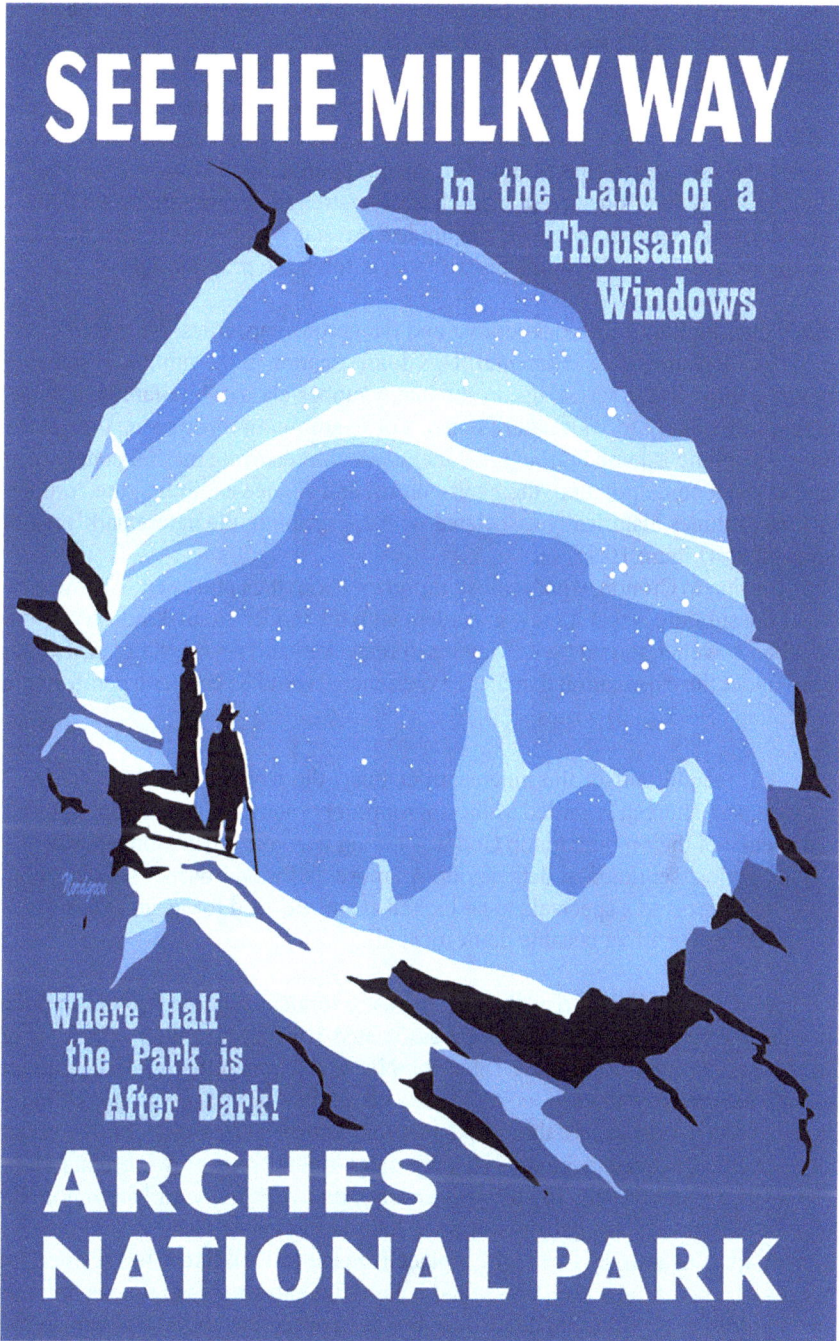

Figure 16.3 Arches National Park. Poster by Dr Tyler Nordgren.

more dangerous version of the astronomical sublime was summed up nicely by Edmund Burke (2018) when he wrote that in

> utter darkness we are ignorant of the objects that surround us; we may every moment strike against some dangerous obstruction; we may fall down a precipice the first step we take; and if an enemy approach we know not in what quarter to defend ourselves; in such a case strength is no sure protection; wisdom can only act by guess; the boldest are staggered, and he who would pray for nothing else towards his defence, is forced to pray for light.

Rather than highlight these disruptive and dangerous capacities, Nordgren's sublime is gentle, tame, and sentimental. This domestication of the sublime is achieved primarily through the series' use of the happy, upper-class, white family, a demographic most likely to visit nature parks. For example, in most posters the astronomical sublime is tamed through the inclusion of a recurring marital couple, who stand in quiet contemplation, their self-control and silence exhibiting the "proper" spiritual attitudes historically associated with appreciating the nature park's magic and mystery (Figure 16.4).

According to Cronon, what is most important about the nature park's domestication of the sublime is that it grew in tandem with the modern tourist industry. While the urban traveller sought "wild" nature to reawaken and revive their spirits, they did so under the expectation that these experiences would be cosy and comfortable. As Cronon (1995, p. 6) writes,

> By the second half of the nineteenth century, the terrible awe that Wordsworth and Thoreau regarded as the appropriately pious stance to adopt in the presence of their mountaintop God was giving way to a much more comfortable, almost sentimental demeanour. As more and more tourists sought out the wilderness as a spectacle to be looked at and enjoyed for its great beauty, the sublime in effect became domesticated.

In addition to enticing tourists through the astronomical sublime, Nordgren relies on a second recurring romantic trope associated with 20th-century conservation movements: frontier nostalgia. "Those who have celebrated the frontier", Cronon writes, "have almost always looked backward as they did so, mourning an older, simpler, truer world that is about to disappear, forever" (7). No doubt, the recent public interest in reconnecting with dark skies stems from growing fears about their swift disappearance. To curb this erasure, many dark sky parks promote the value of darkness by capitalising on what John Urry (1994) calls "glacial time", a touristic form of nostalgia whereby a place is made to look like it has resisted the ravages of modern time. For Urry, what is important about glacial time is not only how it holds the power to facilitate deep bonds between the tourist and human history but how the tourist misconstrues the past as something that endures eternally. In *Half the Park is After Dark*, Nordgren relies on the myth of the American frontier, a location whose qualities of freedom and uncultivated openness have long

Figure 16.4 Shenandoah National Park. Poster by Dr Tyler Nordgren.

inspired the astronomical imaginary, to create his glacial vision of the starry night. Like the western pioneers in the American frontier myth, who imagined themselves as brave heroes reclaiming their freedom from the clutches of civilisation, Nordgren's contemplative family is pictured as the remnants of a lost culture of adventurers who would prefer to live nostalgically in a pre-industrial golden age than confront the so-called "ends of nature".

The irony of this depiction of the starry night should be immediately apparent. Though dark sky parks and preserves are a recent tourist invention, with some parks receiving certification in the past few years, the astronomical scenes represented by Nordgren are explicitly anti-modern (Dunnett, 2015). Against the ugly artificiality of the urban night, Nordgren offers the viewer his nostalgic time-travelling couple, whose rustic attire (the posters dress them in the style of early 19th-century pioneers) testifies to the night's ability to teleport them back in time. Indeed, everything about the series is permeated with a retro aesthetic, wherein all technological markers of the present are disavowed. For example, the posters never refer to the various optical technologies (digital cameras, high-powered telescopes, expensive star trackers) that typically accompany the contemporary astro-tourist, whose pictures of the stars are immediately distributed across various media networks. Instead, Nordgren purifies the starry night by suggesting that the best and most straightforward way to appreciate the night's beauty is through the naked eye, a "luddite" view of the cosmos that ensures the park's status as an ancient inheritance.

Beauty won't save the starry night

By capitalising on the sublime wonders of pristine darkness, astro-tourism reverses a longstanding moral order. Whereas rural places were often associated with decline and regression, today, possessing an uncontaminated view of the stars gives the countryside a moral advantage over the city's excessive illumination. But can beauty alone save the starry night? For many dark sky advocates, the answer is a resoundingly yes. As Terrel Gallaway (2010, p. 79) writes, "those who have never seen a night sky ablaze with stars will lack sufficient information to judge the magnitude of their loss". For Gallaway, only "the passive enjoyment" of the night will help the public change its mind about the value of darkness. Economic incentives, energy efficiency, and other utilitarian policies might offer short-term solutions to the growing problems of light pollution. However, as Gallaway (ibid., p. 81) observes, "utility bills do not well reflect what is culturally special about dark skies—the beauty that is the night's true comparative advantage. It is hard to imagine, therefore, a long-term solution to light pollution that does not explicitly recognize the fundamental role natural beauty plays in social welfare".

I do not doubt that a sky aglow with myriads of stars possesses aesthetic values that may help convince people to "turn down the lights" and conserve darkness; Gallaway's suggestion that we disregard the night's beauty at our peril, however, fails to acknowledge how aesthetics have been (at least in the context of American environmentalism) the principal means for protecting "wild" nature. Furthermore, as I have shown here, promotional materials for dark sky tourist destinations do not shy away from the night's beauty but are dominated by the nostalgic ideals of romanticism. Thus, the idea that the only way to protect the ancient night is to appeal to its natural beauty is myopic at best.

The more pertinent question should not be how can we increase public awareness about the night's beauty, but how does visualising the night through romantic ideals hurt the movement to protect the starry sky? As Willian Cronon (1995, p. 12) shows, the main lesson we can draw from the cultural baggage of "pristine wilderness" is that when environmental conservation is framed around the loss of beauty, it tends to reproduce dangerous ideas about nature that pose a "serious threat to responsible environmentalism". Indeed, as much ecological research shows (Holt, 2012), the standard ethical view that environmental problems are best served by replacing the consumer values of urban society with ones based around unspoiled landscapes or pre-modern times is rarely successful.

This lesson in the dangers of placing too much faith in beautiful places was present at the very beginning of the national park movement. For example, the philosopher John Dewey (1999, p. 101) referred to the public's lust for picturesque locations as an expression of escape, as people used mass travel to flee the most pressing concerns of the industrial age. "Many American critics of the present scene are engaged in devising modes of escape", Dewey observed. "Some flee to Paris or Florence; others take flight in their imagination to India, Athens, the Middle Ages or the American age of Emerson, Thoreau, and Melville. Flight is solution by evasion". While Dewey recognised the beauty and charm of America's newly consecrated national parks, he was adamant that beauty alone could not solve the massive problems brought about by industrialisation. "There is no need to deny the grace and beauty of some of these constructions", Dewey (ibid., p. 102) wrote. "But when their imaginary character is once made apparent, it is futile to suppose that men can go on living and sustaining life by them".

Although Dewey's criticism was made in the early 20th century, when mass travel was in its infancy, much of his observations about the problem of using idealised places to confront contemporary problems are just as apt today. Simone Abram (2003), for example, updates many of Dewey's concerns through her idea of the rural gaze, the idea that urbanites romanticise the countryside by idealising the past through rustic and pastoral touristic images. Like Dewey, Abram is wary of how sentimental yearnings for the picturesque conceals the inequalities, conflicts, and tensions facing rural peoples in the present. As she (2003, p. 35) writes, "the tourist gaze upon the rural landscape is one and the same as the rural gaze which aestheticizes land uses in a nostalgic way in an attempt to distance it from contemporary capital and globalizing processes". Similarly, for Cronon (1995, p. 10) the problem with nature parks is that by defining "pristine" nature as that which lies beyond human intervention, conservationists create a series of binary divisions that erase history. He continues,

In virtually all of its manifestations, wilderness represents a flight from history. Seen as the original garden, it is a place outside of time, from which human beings had to be ejected before the fallen world of history could properly begin. Seen as the frontier, it is a savage world at the dawn of civilization, whose transformation represents the very beginning of the national historical epic.

Dewey, Abram, and Cronon all share the belief that confronting contemporary problems like urban light pollution requires a mode of thinking free from the reactionary and over-simplified impulses of romanticised nostalgia. Instead of enticing viewers with rare and majestic views of the night sky, it is more helpful to interrogate the myths that sustain the idea that "pristine" darkness provides moral fodder for restoring humanity's connection to the night. Nordgren's nostalgic and sublime account of the starry night is undoubtedly astonishing. However, his presentation of the night is neither natural or authentic but an act of rhetorical mythmaking, since what he chooses to show and not to show shapes how tourists experience and understand darkness. As Lawrence Prelli (2006, p. 24) argues, the rhetorical power of images resides as much in what they conceal as what they put on display. "Whatever is revealed through display", Prelli observes, "simultaneously conceals alternative possibilities; therein is display's rhetorical dimension".

One glaring problem concealed by Nordgren's construction of the astronomical sublime is the ongoing material transformations taking place in outer space. While the night sky could be seen as a wild, untouched frontier before the launch of Sputnik in 1957, the contemporary sky continues to be subject to various ongoing technological, political, and corporate expansion projects. According to Peter L. Hays (2015), we now live in a fourth stage of space power, a new space race wherein the elite extract wealth from new forms of space tourism and small satellites. However, despite the global reach of this new stage of colonial space power (some estimate that by 2030 the small satellite industry will block out the view of most stars), most citizens fail to notice the space infrastructure underpinning modern life. As Patrick McCray (2021) notes,

> As I write . . . thousands of objects of varying size are orbiting the earth. Hundreds of these are functioning satellites. These objects girding the globe are critical links in a modern technological and scientific infrastructure that most people reflect on little, if not at all.

Ironically, it is the astro-tourist who is best suited to take notice of the ill effects space expansion projects have on the starry night. Anyone who has tried to capture an image of the night sky knows all too well that satellite trails (bright, thin lines that appear in long-exposure images) are a new visual and material component of the contemporary night sky. Furthermore, even the darkest, most remote dark parks and preserves cannot mitigate this new form of light pollution. Yet by endorsing the idea that national parks offer an untainted connection to the wonders of the ancient night, Nordgren's posters create what Nicholas Mirzoef (2014, p. 217) calls "Anthropocene visuality", a way of imagining environmental problems that "allows us to move on, to see nothing". Thus, one of the dangers of holding onto the idea that the untainted night is our ancient home is that "we give ourselves permission to evade responsibility for the lives we actually lead. We inhabit civilization while holding some part of ourselves—what we imagine to be the most precious part—aloof from its entanglements" (Cronon, 1995, p. 11).

Conclusion

In a recent study published by *Science* (Mortillaro, 2023), scientists used data from the Globe at Night outreach program to determine the annual rate of light pollution. Not surprisingly, they found that global light output had increased between 7 and 10% annually over the past 12 years. To put this loss of darkness in perspective, whereas a stargazer 12 years ago could identify 250 stars in the night sky, today, that same onlooker would see only 100.

For many dark sky advocates, the growing light pollution problem can be solved by appealing to the public's aesthetic sensibilities. Yet when viewed from the history of national park conservation, the idea that beauty will save the night falls short, its fetishisation of the ancient night repeating many of the ideological shortcomings of the myth of "pristine" nature. For when "pristine" darkness is the standard by which we judge the starry night, we not only set too high the standard for what counts as naturally wondrous, but other less pure experiences with darkness are discounted (everyday encounters with gloom that may invite people to care about darkness). More troubling is how the night's romanticisation erases vital traces of modern life, a flight from history concealing the global processes radically altering the night.

To save the starry night, what is needed is an alternative viewing practice that at once acknowledges the night's beauty and brings into focus the hidden technological infrastructures colonising the night sky. One such viewing tradition can be found in the Moonwatchers collective from the first space age. After the Soviet launch of Sputnik, a small group of citizen scientists looked up at the sky each night in search of critical information about the first satellites. Although their astronomical interests were forged in an age of optimism about the benefits of space exploration, a belief in the progress of science that has waned today, their global efforts to bring visibility to a new range of technological artefacts is instructive. For if the role of the critic is to show the gulf between the idealised and the material real (Berger, 2008), the Moonwatchers demonstrate that despite the remoteness of outer space (a disconnect that has helped support troubling and wild idealisations about the night sky), they were able to monitor and track the material infrastructures reconfiguring the night sky.

Luckily, for those interested in bringing back to earth the hidden space infrastructure that contaminates the night, there are options that the astro-tourism industry can incorporate into their visual practises. The space archaeology of Alice Gorman (2019) and the infrastructural media studies of Lisa Parks (2005), for example, are two prominent frameworks whereby images of satellites are the starting point for detailed discussions of the power arrangements and inequalities associated with satellite technologies. Indeed, by starting with a "footprint" analysis of the sky, which foregrounds the material real over the romantic sublime, both ensure that "the earth and the sky above are filled not just with the mysteries and magic of nature, but also with the complex interventions of humankind" (Pasternak and Thompson, 2012).

References

Abram, S. (2003) 'Gazing on rurality', in Cloke, P. (ed.) *Country visions*. Harlow: Pearson, pp. 32–48.

Albers, P. C. and James, W. R. (1988) 'Travel photography: A methodological approach.' *Annals of Tourism Research*, 15(1), pp. 134–158.

Barthes, R. (1999) 'The rhetoric of the image', in Evans, J. and Hall, S. (eds.) *Visual culture: The reader*. London: Sage, pp. 33–40.

Beauregard, R. (2002) *Voices of decline: The postwar fate of US cities*. London: Routledge.

Berger, J. (2008) *Ways of seeing*. London: Penguin Classics.

Blundell, E., Schaffer, V. and Moyle, B. D. (2020) 'Dark sky tourism and the sustainability of regional tourism destinations', *Tourism Recreation Research*, 45(4), pp. 549–556.

Bogard, P. (2014) *The end of night: Searching for natural darkness in an age of artificial light*. New York, NY: Back Bay Books.

Brown, P. L. (2000) *Megaliths, myths and men: An introduction to Astro-archaeology*. Garden City, NY: Dover Publications.

Burke, E. (2018) *The works of Edmund Burke volume 1*. Kindle edn. Berlin: Verlag.

Conn, S. (2014) *Americans against the city: Anti-urbanism in the twentieth century*. Oxford: Oxford University Press.

Cronon, W. (1995) 'The trouble with wilderness: Or, getting back to the wrong nature', in *Uncommon ground: Rethinking the human place in nature*. New York, NY: W.W. Norton & Co., pp. 69–90.

Demars, S. E. (1990) 'Romanticism and American national parks', *Journal of Cultural Geography*, 11(1), pp. 17–24.

Denevan, W. M. (1992) 'The pristine myth: The landscape of the Americas in 1492', *Annals of the Association of American Geographers*, 82(3), pp. 369–385.

Dewey, J. (1999) *Individualism, old and new*. Buffalo, NY: Prometheus.

Donahue, M. Z. (2016) '80 percent of Americans can't see the milky way anymore', *National Geographic*, June. Available at: www.nationalgeographic.com/science/article/milky-wayspacescience#:~:text=But%20a%20new%20atlas%20of,Way%20has%20become%20virtually%20invisible (Accessed: 30 January 2023).

Dunnett, O. (2015) 'Contested landscapes: The moral geographies of light pollution in Britain', *Cultural Geographies*, 22(4), pp. 619–636.

Emerson, R. W. (2012) *Nature and other essays*. Garden City, NY: Dover Publications.

Escario-Sierra, F., Alvarez-Alonso, G., Mosene-Fierro, J. A. and Sanagustin-Fons, V. (2022) 'Sustainable tourism, social and institutional innovation—the paradox of dark sky in astrotourism', *Sustainability*, 14, p. 6419.

Farajirad, A. and Beiki, P. (2015) 'Codification of appropriate strategies to astronomical tourism development', *Applied Mathematics in Engineering, Management and Technology*, 3, pp. 303–312.

Fayos-solà, E., *et al.* (2014) 'Astrotourism: No requiem for meaningful travel', *PASOS Revista de Turismo y Patrimonio Cultural*, 12, pp. 663–671.

Gallaway, T. (2010) 'On light pollution, passive pleasures, and the instrumental value of beauty', *Journal of Economic Issues*, 44(1), pp. 71–88.

Gorman, A. (2019) *Dr Space Junk Vs the universe archaeology and the future*. Cambridge, MA: The MIT Press.

Hays, P. L. (2015) 'Spacepower theory', in Schrogl, K-U. (ed.) *Handbook of space security*. Cham: Springer, pp. 57–79.

Holt, D. (2012) 'Transformation of unsustainable markets constructing sustainable consumption: From ethical values', *The ANNALS of the American Academy of Political and Social Science*, 644, pp. 236–255.

Janowitz, A. (2005) 'What a rich fund of images is treasured up here': Poetic commonplaces of the sublime universe', *Studies in Romanticism*, 44(4), pp. 469–492.

McCray, P. (2021) *Keep watching the skies! The story of operation Moonwatch and the dawn of the space age*. Princeton, NJ: Princeton University Press.

Melbin, M. (1987) *Night as frontier: Colonizing world after dark*. New York, NY: Free Press.

Mirzoeff, N. (2014) 'Visualizing the anthropocene', *Public Culture*, 26(2(73)), pp. 213–232.

Mitchell, D. and Gallaway, T. (2019) 'Dark sky tourism: Economic impacts on the Colorado plateau economy, USA', *Tourism Review*, 74(4), pp. 930–942.

Mortillaro, N. (2023) 'Goodbye, dark sky. The stars are rapidly disappearing from our night sky', *CBC News*. Available at: www.cbc.ca/news/science/light-pollution-increasing-1.6719034.

Muir, J. (2020) *Our national parks*. Kindle edn. City of Alexandria, VA: Library of Alexandria.

Neumann, M. (1999) *On the rim: Looking for the grand canyon*. Minneapolis, MN: The University of Minnesota Press.

Parks, L. (2005) *Cultures in orbit*. Durham, NC: Duke University Press.

Pasternak, A. and Thompson, N. (2012) 'Forward', in *The last pictures*. Berkeley, CA: University of California Press.

Patin, T. (ed.) (2012) *Observation points: The visual poetics of national parks*. Minneapolis, MN: The University of Minnesota Press.

Prelli, L. J. (2006) 'Rhetorics of display: An introduction', in Prelli, L. J. and Benson, T. W. (eds.) *Rhetorics of display*. Columbia, SC: University of South Carolina Press, pp. 1–40.

Silver, D. A. and Hickey, G. M. (2020) 'Managing light pollution through dark sky areas: Learning from the world's first dark sky preserve', *Journal of Environmental Planning and Management*, 63(14), pp. 2627–2645.

Tennyson, A. L. (1987) 'God and the universe', in Ricks, C. (ed.) *The poems of Tennyson in three volumes*. 2nd edn. Harlow: Longman.

Urry, J. (1994) 'Time, leisure and social identity', *Time & Society*, 3(2), pp. 131–149.

Van Dyke, H. (2005) *The poems of Henry van Dyke*. London: The Gutenberg Project.

Virilio, P. (2000) *A landscape of events*. Translated by Rose, J. Cambridge, MA: The MIT Press.

Ward, S. (2005) *Selling places: The marketing and promotion of towns and cities 1850–2000*. London: Routledge.

Weaver, D. (2011) 'Celestial ecotourism: New horizons in nature-based tourism', *Journal of Ecotourism*, 10(1), pp. 38–45.

Part 7

Conclusion

17 Under the night

The futures of dark skies

Nick Dunn and Tim Edensor

Introduction

Throughout this book, we have encountered numerous ways in which dark skies are conceived, encountered, perceived and form communities around them. The approaches collected here represent a significant cross-section of the diverse inquiries concerning dark skies and the various entanglements they coproduce across time and space. The capacity of the arts, humanities, and social sciences to offer fresh perspectives and critical accounts of dark skies is evident in the different chapters compiled in this book. In addition to the new understandings they individually provide, collectively they suggest ways in which we might form a more holistic, relational, and nuanced knowledge of dark skies. This is important and urgent. Dark skies are under threat. The exponential growth of light pollution, coupled with a lack of awareness and understanding of its negative impacts, has led to profound social, environmental, and health consequences and critically, greatly diminished access to dark skies. Pivotal to addressing these issues is the need to form a more diverse and subtle understandings of darkness itself (Dunn and Edensor, 2020). Since darkness is multiple, situated, and contested (Morris, 2011), so too are the manifold ways in which its presence is debated in both positive and inimical ways.

The associations and values attributed to dark skies are now under wider exploration through different academic, aesthetic, creative, ecological, and socially inspired reappraisals. As knowledge about our impact as a species upon the planet grows, we are in danger of losing connection with the very source of darkness that has been with us and our countless ancestors since human beings first evolved and indeed, long before that. Dark night skies are essential to the rhythms of life on Earth, with the evolution of living things stemming from our planet's rotation around its axis as it revolves around the sun. Dark skies are vital to the health of our planet, its ecosystems, and its many inhabitants, including humans. This problem is no longer confined to what happens on the ground. As chapters by Neha Khetrapal and Dwayne Avery emphasise, the proliferation of satellites and space debris in orbit around the Earth is also impacting on the night sky. Initially, this was thought to be increasing the overall brightness of the sky at night (Kocifaj et al., 2021), threatening ground-based astronomy as well as a diversity of stakeholders and

DOI: 10.4324/9781003408444-24
This chapter has been made available under a CC-BY-NC-ND 4.0 license

ecosystems reliant on dark skies. More recent research indicates that this type of light pollution will have major implications for the loss of astronomical data (Barentine *et al.*, 2023), having direct consequences for our knowledge of the world around us and the universe beyond.

Dark skies are also an integral wellspring of inspiration, providing a powerful connection to nature and offering opportunities for recreation. As we have emphasised, this is leading to an extinction of experience (Soga and Gaston, 2016) amongst humans. In some cultures, sky traditions remain an important aspect of their social customs, cultural traditions, and religious beliefs but are threatened. Accordingly, dark skies represent a form of heritage that deserves commitment to their preservation and safeguarding for future generations of humans and non-humans.

The chapters included in this book provide numerous examples of interdisciplinary inquiry that seek to elucidate on specific experiences of our relationships with dark skies and underlines that this a key more-than-human relationship of which we are a part. The diverse approaches here accentuate the importance of acknowledging the plural and located aspects of dark skies. We relate to the sky at night from specific locations, and so the surrounding environment and the weather conditions particular to that time directly inform our experience of the night sky, whether as one of darkness resplendent with stars and other celestial bodies or as one with circumstances that obscure our access to such encounters. Embracing the diversity, plurality and situated qualities of places within which dark skies may be experienced, the contributions to this book collectively offer much-needed alternatives to prevailing conceptions and open up new trajectories for future research, as we identify in the following sections of this chapter.

We focus on four key themes that have emerged. First, we identify some of the diverse geographies and histories through which dark skies have been conceptualised and experienced. Second, we explore how the quest for dark skies and their management is motivating the broader reconfiguration of the relationship between light and dark. Third, we explore how the chapters in this book reinforce the need to hasten the reconnection of people with dark skies, and we identify the key processes of reattunement, re-enchantment, and redistribution of the nocturnally sensible as processes through which this might be accelerated. We conclude by emphasising that this also requires the development of inclusive dark sky communities from the local to the global so as to incorporate the diverse approaches and acknowledge the divergent desires that dark sky sites fulfil.

Multiple histories and geographies of dark skies

The relationship between dark skies and human beings is a story of manifold narratives, reflecting the diverse histories and geographies of their role in cultural associations, social traditions, religious beliefs, and scientific understanding. The various connections with dark skies across time and space are providing new perspectives that widen our knowledge and understanding of their role in different cultures throughout histories as we discuss in our introductory chapter. The multifarious engagements that humans have with dark skies span historical, artistic, conceptual, and sensory modes of connection. That some of these practices and their values have endured while others have altered in relation to shifting contexts enables us to view the past as a series of domains, many that lie outside of the dominant boundaries of Western thought. Moreover, the diversity of relationships with dark skies is not limited to humans, with many non-humans having vital connections to the nocturnal rhythms of the planet and the essential qualities of natural darkness that can enable them to flourish or perish.

In examining one of the many historical contexts through which people have engaged with dark skies, Neha Khetrapal's inquiry into ancient Indian civilisation illustrates that pervasive contemporary negative perceptions of the night emerge from specific historical and cultural contexts; they are not reflective of wider cultural and social practices that took place after dark. These beliefs, rituals, and imaginaries continue to contribute to cultural development, but critically, they also foster growing awareness that we require new tools and paradigms for conducting research into the spatiotemporal nocturnal dimensions of human life. By tracing the relationship with dark skies from ancient celestial observations, Khetrapal investigates how night aesthetics and imaginaries are transferred across generations through material forms while underscoring the loss of both as a result of changing attitudes towards the night. Setting out the contemporary implications for reconnecting with dark skies, she pleads for a nocturnal anthropology that acknowledges this as a form of heritage worth preserving, for it will contribute to attempts to challenge the multiple problems caused by excessive illumination.

The interpretations, values, feelings, and practices that surround dark skies in present times are similarly variable and yet distinctively modern imaginations continue to be dominated by the idea that cities are supposed to be bright. This is compounded through the widespread and rapid growth of urbanisation that has obstructed our aesthetic and imaginary connection to dark skies. Yet the definition of a dark sky is fuzzy because of the circumstances in which it is beheld. Yee-Man Lam's research shows that more broadly, depictions of a "dark sky" paradoxically include both the surrounding earthly darkness that permits us to see the stars of the night sky and the pervasive darkness that conceals such celestial views. More specifically, she examines the term "dark sky" in the context of Hong Kong and across both its Chinese-speaking and English-speaking cultures. Hong Kong is widely cited as one of the most light-polluted cities in the world, and a key finding is that there is no equivalent to "dark sky" in the Chinese-speaking context, only a "night sky", which is neither dark nor starry. This absence may seem to be a simple one of nomenclature, but it perhaps reflects a symbolic and more problematic issue—that the darkness here is no longer present to be appreciated and we need to reconnect with gloom rather than accept an illuminated sky as an instrumental backdrop to urban life. In assessing the global spread, ubiquitous in many geographical contexts, of the foregrounding of illumination as integral to nocturnal experience, the scale of the challenge to reappraise darkness and dark skies is evident.

Yet the divergence of values, interpretations, and experiences of darkness underline that geography and history are important in grasping the effects and qualities of dark skies, further accentuating our claim that darkness is multiple and situational.

In this book, amongst other settings, we learn of the diverse dark skies of ancient India, Essex, Lancashire, Cumbria, Mayo, the Welsh Borders and Snowdonia, New Zealand, Canada, the USA, Germany, and Hong Kong. And we venture into designated dark sky places, woods and forests, urban metropolises, English and American national parks, bird reserves, Irish and Welsh dark sky towns, borderlands, Indian temple complexes, and megalopolises.

We underscore this point to move away from any essentialist conceptions of darkness and dark skies; rather we celebrate their multiplicity. Critical in conditioning the experience of darkness are the different experiences of the diurnal and seasonal distribution of sunlight and the angle at which the sun hits place in accordance with latitudinal, longitudinal, seasonal, and environmental contexts. They shape the conditions of twilight, midnight, deep night and dawn, and many minor stages in-between. This is exemplified by the very different expressions of twilight that can be experienced in different realms. Peter Davidson's (2015) highly nuanced account of the subtle, ever-changing, and manifold sensory and metaphorical potency of twilight particularly captures the shifting experience of gloom as the sun goes down in contexts that range across different latitudes and landscapes. Davidson depicts the shifting hues and emerging shadows with thick, poetic descriptions while drawing on a wealth of historical, literary, and artistic representations to explore the sometimes intoxicating, sometimes melancholic affective and sensory impacts of the disappearing sunlight and the emerging darkness. Other writers also seek to capture these distinctive, evanescent moments. The accelerated speed of the advent of sunset in Aotearoa is captured by Lees (2022, p. 29), who describes the sun setting "at a low angle, filtered through the hazy dust and water droplets" of the atmosphere. Rather differently, Durand (2021) notes the luminous blue hour in which water and sky seem to merge that is celebrated across northern countries as a mystical, divine, and aesthetic time. Yates (2012, p. 143) focuses on the hugely diverse appearances and atmospheres of daybreak that emerge following his nocturnal ramblings, contrasting the winter's "brief blossoming of gaslight blue before the sun comes up" with a June dawn that "seems to begin as soon as the evening twilight tapers to a fingertip along the northern horizon". We might also identify the non-visual sensations that foretell of the end of darkness, sonically, for instance, by the distinctively situated birdsongs of the dawn chorus that herald the advent of daybreak or the clinking sounds of the milk delivery in the early morning hours of the United Kingdom.

A further factor in shaping the experience of dark skies and darkness are the divergent affordances of the landscape. The shadows of the forest conceal the light from the sky and generate a deep blackness under their canopies, pools of water that stretch across moorland glimmer and reflect skies, and the looming shapes of mountains obliterate the moon. Islands also offer distinctive experiences of dark skies, as Dixe Wills (2015) describes in his visit to the Scottish island of Coll, recently designated a dark sky community, and as Ada Blair (2016) explores in her study of the similarly designated small English Channel island of Sark, which, she asserts, offers positive experiences of wellbeing, communality, a return to childlike sensations, and spiritual enhancement for residents and visitors.

Particular landscapes remain saturated with mythic associations when darkness falls—marshes, moors, old woods, deep riverine valleys, megaliths, stately homes, and ruins of diverse age are replete with intimations of the uncanny and the haunted. The imposing rock silhouettes of Iceland assume forms that are held to resemble petrified trolls. Wills (2015) provides an example of such storied landscapes, spending a night at the edge of myth-saturated Dartmoor's ancient Wistman's Wood. With its stunted, twisted oak trees, after night falls it has been imagined as a misshapen arboreal realm that hosts devil dogs and pixies.

As in Hong Kong, visual experience of dark skies is also variously conditioned by levels of artificial illumination. James Attlee (2011, p. 3) contends that only in "the great deserts and oceans" can the moon be fully appreciated under dark conditions, while Paul Bogard (2013) witnesses star-filled skies in the vast blackness of Death Valley, California, although he finds that other formerly crepuscular American national parks have been invaded by light pollution. Across the earth, we can still point to Chile's vast Atacama Desert, deep-sea trenches, and caves that remain largely devoid of light. In many northern settings, the pervading darkness of the polar night also replaces the baleful effects of over-illumination. By contrast, Judy Spark (2021) describes the experience of spending the night in a camping ground on the edge of large city, "just out of the orange city fug, just on the edge of the darker countryside". The urban skyglow radiates nearby but the city's lights do not fall on the rural spaces close at hand, generating a peculiar sense of detachment.

Reconfiguring the relationship between light and dark

The encroachment of skyglow testifies to the serious problems of over-illumination across the world that many chapters discuss in this book in pointing to the demise of dark skies. One key consequence is that dark and light remain prominent dualities in many framings. These oppositional dynamics are complex, but typically, they can reinforce power relationships, persistent cultural associations, and enduring social values. Western in origin, these binaries are rooted in regulatory ideas about diminishing oppression and violence yet also tenaciously express popular notions that light symbolises clarity and coherence.

This spatial expansion of light is not recent by any means. Between the 16th and 18th centuries in Europe, changes in attitudes and beliefs towards the night were especially significant in shaping perceptions of darkness that largely endure to the present day (Koslofsky, 2011). These societal transformations prompted new opportunities for leisure and labour requirements, which, in tandem with technological developments in artificial lighting, reframed the night as an extension of the day, fundamentally altering many of the positive connotations of darkness that formerly existed. This binary situation continues to be strengthened by the ongoing colonisation of the night (Melbin, 1978), which continues to render darkness as synonymous with regression, impoverishment, or unproductive time (Crary, 2013). Yet many subaltern and alternative positive evaluations of darkness have always challenged these orthodoxies; indeed, in present times, paradoxically, as illumination is frequently used to extend visibility and surveillance and light pollution is rapidly increasing, darkness is being promoted and savoured as a luxury (Beddington, 2023).

It is evident from the growing body of scientific research that most of the problems caused by light pollution are preventable. Globally, light emissions represent 246,238 gigawatt hours of energy which is 1% of global emissions (Sánchez de Miguel *et al.*, 2021). Amidst the current ecological crisis, it is easy to overlook the need for natural darkness since, compared to climate change, acid rain, exotic species, and habitat destruction, it can seem less urgent. Yet this underestimates the role of the rhythms of light and dark in supporting the planet in three significant ways. Firstly, they enable the primary productivity of food webs, producing oxygen and regulating carbon cycling and sequestration. Secondly, they encourage more and greater diversity of interactions between species. Thirdly, they help ensure ecosystem resilience.

Recent research reinforces the negative impacts light pollution is having upon human health (Brown *et al.*, 2022) and that anthropogenic activity, especially light pollution, is having on non-humans (Gaston, Gardner and Cox, 2023). Besides contributing to the increase in carbon dioxide in the atmosphere, light pollution disrupts the natural patterns of humans and non-humans. The ecological impact of artificial light at night on influencing the behaviours, physiologies and ecosystems of non-humans has been well-demonstrated. Yet the potential of light pollution to alter plant phenology and its corresponding cascading effects on herbivore is less clear, though there appears to be correlation: the relationship between light pollution and plants has thus yet to be proved (Ffrench-Constant *et al.*, 2016) and requires greater research.

Therésa Jones and Marty Lockett highlight the vital role of light/dark rhythms in shaping daily, monthly, and seasonal cycles before explaining how artificial light at night disrupts these natural cycles. They emphasise the harmful link between the presence of light at night with shifts in growth, development, and survival in non-human animals, alongside the suite of health consequences that affect all animals, including humans, such as increased risk of disease, loss of immune function, reduced fertility, disrupted sleep, weight gain, and (in humans) impaired mental health. Jones and Lockett then discuss the conflict between human desire and decision-making power for nocturnal brightness and offer suggestions for better balance of light and dark through mitigation and resolution that would dramatically benefit all life on Earth. Building upon this theme, Kimberly Dill presents an environmental ethical framework to analyse how the loss of natural darkness negatively impacts upon the conservation and restoration of biodiverse forested ecosystems while simultaneously damaging the relationships that humans form with them. She explores how the drivers of ecological change, including dramatic increases in light pollution, have measurable effects on network resilience. Dill illustrates the interdependencies between plants and pollinators, with the loss of one having significant, even irreplaceable, consequences on the flourishing of the other. Embedded in her analysis is an implicit gesture toward the malign climate impacts of light pollution. In particular, she reminds us that forests function as carbon sinks for the sequestering of atmospheric carbon dioxide into the trees, grasses, shrubs, and mycelial networks that compose them. Hindering the regeneration of forests, which under naturally dark conditions could serve as effective carbon sinks, reinforces the need for the preservation of natural darkness as fundamental to forest, human, and global health.

A major challenge in combatting the rise in light pollution and the resultant disappearance of dark skies concerns our awareness about the loss, for we have largely become unreflexively habituated to levels of light that would have been previously unimaginable. It is difficult to recognise how much darker the nights once were, even during our own lifetimes, due to how light pollution increases and spreads, illustrating the insidious processes through which it has become seemingly inseparable from a modern and globalised world (Dunn, 2019).

Yet given the destructive environmental effects resulting from the continuing spread of illumination across dark spaces, new thinking amongst policymakers, environmentalists, technologists, and light designers is imperative. Dark sky places have been exemplary venues for reconfiguring the relationship between light and dark, providing ideal settings for the creation and installation of alternative, responsible, more sustainable forms of artificial illumination. The benefits of this are twofold. First, the design and implementation of different lighting practices promotes accessibility, drawing in a wider array of people than who would normally visit such locations, thereby enhancing our knowledge and extending experiences of dark sky places. Second, these alternative lighting installations can serve as prototypes for a broader rethinking and reconfiguring of our relationship with light and dark beyond these specific settings. They provoke us to reimagine how our world might be if there was a more nuanced and balanced relationship with artificial light after dark.

The appetite for responsible lighting and alternative approaches to designing with darkness in ways that produce more varied, sustainable, and aesthetically pleasing nocturnal environments is growing (Dunn, 2020; Lowe and Rafael, 2020; Edensor, 2022; Zielinska-Dabkowska, 2022). In their chapter, Tim Edensor and Dan Oakley reconsider the role of illuminations in festivals within dark sky locations, setting out how we might approach creative ways to enjoy such interventions while supporting the fundamental aspect of the dark as integral to these experiences. They discuss how conflicts may arise with the introduction of inappropriate artificial lighting into places of designated dark sky status and call for more influential guidance for light festival organisers and technicians to deploy less intrusive lighting to enhance conservation and protection. Edensor and Oakley then present a series of more progressive approaches to engaging with darkness by privileging sustainable lighting that minimises the impact on living beings at night besides creatively enhancing the aesthetic, historical, and cultural characteristics of nocturnal place and maintaining the ambience of darkness. Kerem Asfuroglu, meanwhile, draws upon his extensive experience as a lighting designer working with several dark sky communities in the UK and Ireland. His chapter highlights the importance of specific, placed-based aesthetics and material qualities, environmental imperatives, social agents, and political contexts in providing guidance when devising lighting schemes for dark sky places. His three innovative case studies are emblematic of the much-needed transition toward more sustainable, aesthetic, convivial, and place-specific lighting design and implementation.

Asfuroglu's enthusiasm for designing for dark sky places chimes with the broader revaluing of darkness that is being more substantively taken up by professional lighting designers and articulated through their organisations. One important group, LUCI (Lighting Urban Community International), identify seven goals in their *Declaration for the Future of Urban Lighting* (Ross *et al.*, 2023). Under the goal of embracing net zero lighting, they emphasise that members should "[a]pply design and planning approaches that enable us to achieve more with less light. Such approaches include: developing sustainable lighting master plans, preventing excessive private outdoor lighting, glare, and using dimming strategies". They call for a general reduction in brightness and in the numbers of luminaires deployed and for the creation of dark infrastructures, such as unilluminated wildlife corridors. They specifically draw attention to commercial lighting, a great contributor to light pollution that, they argue, should be more tightly regulated. The second goal, "Minimising Light Pollution for All Living Entities", calls for a reduction in environmentally harmful bluish-white LEDs. By also focusing on light festivals in their declaration under the goal of "Harnessing the Transformative Potential of Light Art", they contend that such occasions offer experimental possibilities for future lighting that might open up new experiences of urban space. Further, they ask how the development of ecologically oriented approaches to illumination in festivals may act educatively in foregrounding questions about the balance between light and dark, and they call for greater participation from local actors while championing place-based and community-driven approaches.

Such perspectives have also been adopted by local authorities; in Eindhoven, a Lighting Master Plan specifies that "a respect for darkness is a key tenet", with the most immediate effect of this maxim a reduction in illuminated advertising. Influential light designer Roger Narboni (2017, p. 51) has pleaded for a "nocturnal urbanism" that protects and preserves darkness and supports green spaces and blue areas such as parks, canals, and rivers by reducing illumination. Similarly, the lighting design firm Concepto (2012), at which Narboni works, have developed a master plan for Rennes, France, that introduces *dark zones* into the city core. And as Claire Downey (2021) reports, in Paris "public lighting contracts (including street lighting) are set to expire in 2021, entailing a wholesale rethinking of how the city is lit", a process that will seek to establish a different balance between light and dark, where the latter is not conceived as that which must be extinguished. In the UK too, Dark Source (2020) have initiated a number of projects to install creative and ecocentric lighting that span dense urban environments, small villages, and dark sky parks. The inspirations developed by these designers and artists suggest that there is enormous potential for devising approaches in dark sky parks that either sidestep the use of illumination entirely, maintaining darkness, or deploy it in subtle, minimal ways to accentuate gloomy qualities. It seems that light designers too are conceptually grasping darkness as a continuum composed of diverse shades, intensities, and depths rather than a singular blackness and are seeking to encourage people to become attuned to its multiplicities.

In his chapter, Taylor Stone argues that solutions to dark skies should be conceived as reconfiguring practices of maintenance and repair. However, while this also embraces the redesign of urban illumination technologies, the emphasis is on a more overarching notion: the repair of our relationship with the urban night and darkness. This should partly be perpetrated, he considers, through processes of rewilding that include humans as integral to urban wildscapes and not separate from them. Stone further maintains that policies should avoid attempts to recreate historical forms of darkness by focusing on the processual, situational contexts for reshaping urban illumination, thereby remaining open to the multiple possibilities that might arise from design and technological innovations.

Reconnecting with dark skies: reattunement, re-enchantment, and redistribution

As we state earlier, the unreflexive, habitual apprehension of nocturnal environments that pulsate with electric light consolidates common-sense understandings that emerge from a modern, safety conscious mode of design that banishes darkness to the margins. Accordingly, though dark skies offer many obvious benefits, this remains a significant obstacle in their appreciation and engagement. One key element to overcoming this barrier is through the curation and staging of encounters with darkness through creative work that conveys the multiple nuances of its astronomical, environmental, symbolic, and aesthetic aspects. Creative engagements that help people reconnect with the different characteristics, forms, and temporalities of darkness, and where and how it might be experienced divergently, are vital in opening up new dialogues and elucidating fresh perspectives. The role of artistic and creative practices to enrich theories and methods can explore how dark places might be imagined and conceived through alternative ways of environmental, cultural, and spatial knowing.

In her investigation of Galloway Forest Dark Skies Park and the Northumberland Dark Skies Park, Ysanne Holt attunes us to the potentialities of this borderland region's dark skies as a means of thinking across and beyond borders and boundaries, underscoring the ecological sense of the interrelations of humans and non-humans and of material and immaterial forces. Holt identifies these dark sky places as providing fertile, deliberative spaces within which we can imagine future alternatives, better forms of connectivity, and a stronger embodied awareness of our own existence in fragile and vulnerable environments. She demonstrates the wider relevance and implications of interconnected ways of thinking, being, and doing within and across sites, fields, borders, and boundaries, and supports the broader need to develop alternative, future-oriented forms of placemaking in response to the challenges of climate change.

The fragility of the environment is a theme that also emerges in the reflective account of Louise Beer who examines how living under the dark skies of Aotearoa New Zealand has influenced her artistic and curatorial practice and how her artwork invites onlookers to explore their own changing relationship with the night. Through her work, she demonstrates how representations of dark skies and light-polluted skies can resonate with notions of the climate crisis and deep time. Much of her work recognises grief as a vital aspect of the human condition, that is once highly personal and universal. The loss of dark skies, therefore, is positioned as part of our ongoing anguish about the climate crisis as our fundamental connection to the night sky disappears. While reflecting on her personal understanding of life, death, darkness, and light, Beer's artistic practice invites the audience to explore their own changing relationship with the night.

Rather differently, Ellen Jeffrey's research examines the concept of nightfall as the journey of a place towards shadow and darkness. As a time of transition and heightened sensitivity, she draws upon her practice as a dance artist to explore what we can learn of the nocturnal world by dancing in the dark. Jeffrey considers whether by conserving the night's darkness we might also conserve ways of moving through the world. The ways we kinaesthetically relate to place are particularly resonant in the dark where the imagination anticipates form and space rather than recognising it. Meandering, wandering, and slowing down are vital aspects of this embodied approach, and these are reappraised and proposed as essential to our encountering and understanding of the more-than-human environments we live with. By engaging with night as a world in which what is visually perceived no longer equates to clarity and accuracy, her investigation posits the dancer's movement in the dark as a patterning of sense making and making sense of the temporalities of a nightscape.

How can we capture the nuance and value of darkness and to what end? Rupert Griffiths, Nick Dunn, and Élisabeth de Bézenac's inquiry into different ways in which depictions of darkness can be collected and combined seeks to create a thick description of dark places. Drawing on fieldwork that utilises unattended sensors, photography, and walking, evocations of darkness are presented that reflect systematic environmental observation, imaginative interpretation, and bodily rhythms and sensation, respectively. Through bringing these practices together, equal value is placed upon on the material and imaginary dimensions of the night and darkness. They discuss how such situated understandings of place grounded in the lived experience of humans and non-humans might be applied to inform urban design and policy strategies that consider the urban environment as a more-than-human ecology. Griffiths, Dunn, and de Bézenac argue that being able to capture different registers of the nocturnal environment can convincingly inform how design conceives and responds to the coproduction of place after dark. Such approaches will be imperative if we are to find effective ways of regulating light pollution through policy (Morgan-Taylor, 2023) and developing appropriate technological solutions to address its ongoing deleterious effects on other species and ecosystems (Jägerbrand and Spoelstra, 2023).

Holt, Jeffrey, Beer, Griffiths et al., and McGhie and Marr reveal how artistic installations and collaborative practices in dark sky places need not rely on any illumination at all, drawing out the experiential possibilities for multisensory experience. In the introduction, we identify how a wealth of activities in the dark, including eating, listening, touching, walking, and moving in other numerous ways bring out such sensory engagements, in addition to stimulating the visual apprehension of dark space in new ways. Such experiences can undercut the normative sensing of the world and enrich the sensory experience of what initially feels strange: entering the dark worlds that emerge after night falls and reconfiguring our apprehension of the world. In his contribution to this book, Taylor Stone argues that new stories about urban darkness need to be narrated that help us reimagine how dark environments might be considered, interpreted, practised, and designed otherwise. Such alternatives will also be vital, and artistic endeavours, immersive movement, light design and landscape architecture in dark sky places offer potentialities for redistributing the nocturnal sensible.

In this context, a consideration about the unequal distribution of light and dark, and the values, interpretations and feelings that surround them, is usefully contextualised by Jacques Rancière (2009, p. 13), who explains how both sensory experience and making sense are inherently political since they concern "what is seen and what can be said about it, around who has the ability to see and the talent to speak, around the properties of spaces and the possibilities of time". He claims that such regimes of the sensible are shaped by particular values espoused by the powerful, who are able to configure environments as common sensical realms that are difficult to imagine otherwise. The production of nocturnal space and the creation of forms of illumination and darkness that appear to be part of the way things are and should be constitute such sensory regimes. This produces the illusion that "the configuration of a specific space, the framing of a particular sphere of experience, of objects [is] posited as common and as pertaining to a common decision" (ibid., p. 24). Accordingly, the orchestration of space through material production, design, interpretation, and policing forges a sensual, material realm that demarcates what is seen, smelled, heard, and touched. This produces "spatio-temporal-perceptual complexes that invite and encourage some attentional engagements and inhibit others, that *shape* our attentional performativity" (Hannah, 2013, p. 242) and activate particular affective and sensory responses. According to Matthew Gandy (2017, p. 354), this distribution of the nocturnal sensible is "integral to the changing sensory characteristics of late modernity", its quotidian "affective dynamics", and rhythms and "pervasive atmospheres of distraction", anaesthesia, and predictability (Edensor, 2014).

The present situation is thus one in which we are losing the darkness (Smith *et al.*, 2023), and negative connotations still require unravelling and unpacking to aid our understanding of the potent affective and sensory dimensions of darkness. To facilitate such a process, the works, writings, events, and exploratory mobilities we have discussed, along with many others, can foster new modes of attunement (Brigstocke and Noorani, 2016) that redistributes the aesthetic possibilities of dark places, encouraging us to sense and inhabit nocturnal settings differently, and provoking critical approaches to normative ways of seeing, making and being in the dark. Many contributions to this book reattune us to the sensory and affective pleasures of darkness and dark skies and contribute to a broader project: to profoundly shift the baseline of experience concerning darkness and reshape our values in relation to the protection of dark skies and the environment.

Further, having discussed in our introduction Weber's thesis that modernity heralded progressive disenchantment with the world, we argue that dark sky places offer great potential as realms in which we experience *re-enchantment*. By extending and heightening our sensory capacities through encountering design, movement and stillness, and artistic installations and practices that attune us to dark places, we might experience such re-enchantment. Under such circumstances, Jane Bennett (2001, p. 5) notes, we "notice new colours, discern details previously ignored, hear extraordinary sounds, as familiar landscapes of sense sharpen and intensify". During such experiences we may become "transfixed, spellbound", not only by being charmed and delighted but through the peculiar experience of "being disrupted or torn out of one's default sensory-psychic-intellectual disposition".

Contributing to critical interventions that work to reattune and re-enchant people to darkness while redistributing nocturnal experience are recent events staged at designated dark sky spaces. An expanding array of dark sky festivals feature a compendium of potential activities in which visitors may engage. For instance, Exmoor Dark Skies Festival (2023) stages astrophotography, stargazing, lectures, dining, wildlife walks, ghost walks, night running, storytelling, singalongs, craft workshops, and baking competitions. Another noteworthy series of events took place across the UK in 2022, an extensive programme staged by arts group, Walk the Plank, at 20 different landscapes across the UK. Green Space, Dark Skies (2022) attracted thousands of people to walk through a range of dark sky settings, national parks, and marginal sites. Each participant was furnished with a "geolight" and instructed to act as a "lumenator" via headphones. Filmed above by a drone equipped with a camera, large patterns of light were captured across the landscapes by the formations enacted by the walkers. One outcome was a film broadcast on BBC television that conveyed an impressionistic portrait of these diverse landscapes at various stages of dusk and darkness, of the crowds of participants and the light displays they collectively created that animated the gloomy spaces through which they moved. Walking through crepuscular landscapes, their experience was successively shaped by being immersed in darkness, being part of a large torch-bearing party and performing with these lights to co-create a spectacular temporary artwork. This large-scale programme exemplifies one way in which the redistribution of the nocturnal sensible will require us to reconnect with the nocturnal commons. This requires a wider global community of public and professionals to be involved as the *who* in such processes, as we now discuss.

Conclusion: building dark sky communities

Success in designating and developing dark sky projects resides in a nuanced understanding of their situated natures, bringing together the knowledge and skills of a host of professionals and local people. The communities that form around dark skies are essential to their ongoing preservation and safeguarding of their qualities. These may comprise people and groups who are relatively fixed in terms of geography and have a particular attachment to a locality or workplace, but also include mobile groups that seek to build and sustain their affinity with dark skies according to various motivations and desires.

In this book, we grasp a sense of the expanding array of different communities that testify to the growing diversity of the practices, interests, forms of expertise, and political objectives that surround dark skies. While astronomers and stargazers have been most prominent in promoting dark sky places, these have been supplemented by indigenous communities who seek to preserve and transmit their understandings of the night sky, local authorities and tourism promoters, diverse artistic practitioners, creative non-fiction writers, light designers and festival organisers, ecologists and archaeologists, those performing religious rituals and therapeutic processes, nightwalkers and runners, and local people.

Implicitly, then, this book reflects how we need to incorporate more voices, inputs, and perspectives in developing dark sky places and practices, for understanding the histories, cultures, and practices of darkness reflects the considerable diversity and plurality that is too-often discredited and underrepresented. This inclusive acknowledgement also facilitates a broader conceptual approach. For as with visual scrutiny of places and landscapes during daylight, dark sky spaces cannot be objectively classified according to fixed systems of categorisation; conceptions of landscape are invariably informed by situated, culturally specific modes of scholarly enquiry. There is no essential definition of what constitutes a dark sky, no fixed meanings and values that can be imposed upon how they are interpreted and valued. Greater understandings about historical ways of engaging with dark skies from non-Western and indigenous perspectives, and by groups of different generations, ethnicity and gender, and subaltern and marginality, will lead to moves away from standardised and homogeneous processes and towards a more inclusive range of advocates in arresting the widespread decline of the "nocturnal commons" (Gandy, 2017).

In exemplifying local collaboration, Kerem Asfuroglu emphasises the critical role that highly supportive communities and associated local authorities, both who have a stake in the outcomes of dark sky places, are essential to the success of the lighting schemes he designs with them. Diverse communities may also be engaged through collaborative practices and innovative, experimental, arts-based exercises in dark sky places, as chapters in this book reveal. Asfuroglu's account emphasises that widening participation in dark skies communities is thus a fundamental tenet to reconfiguring the relationship between light and dark. As he explains, besides developing a medley of light technologies and designs to honour darkness, such strategies may require significant local community involvement to assuage concerns about safety, aesthetics, and place-image.

Helen McGhie and Natalie Marr in particular reveal how their respective creative practices as artist-researchers enables them to share insights and challenges of working with partners and using dark sky parks as research sites. They explore how arts-based research can enhance nocturnal recreational activities while inviting stakeholders and different groups of beneficiaries to engage in the co-creation of novel dark sky knowledge and practice. As artist-researchers, they emphasise the co-production of their work with dark sky communities, a process that enriches their practice, applies creative skills to distinctive environments, and responds to recent calls for interdisciplinary approaches to research.

In their chapter, MacMillan et al. similarly place an important emphasis on community involvement but also seek to expand participation through promoting dark sky settings as fertile realms in which to rethink how sustainable tourism is conceived, produced, and managed. Drawing upon examples from Mayo Dark Sky Park in Ireland, they argue for increasing community custodianship and pro-environmental behaviours to support the qualities of places after dark. They stress the urgent need to develop a framework of ethical principles for dark sky tourism to minimise any impact on the environment. They further contend that the symbiotic relationship between darkness as a resource and as an intrinsic cultural asset can only be sustainable if it draws upon the collective and collaborative knowledge of community stewards, practitioners, and agencies.

The design and management of dark sky places and the range of opportunities they offer must bring in diverse perspectives and crucial counterarguments to question commonly perceived attitudes, as Nona Schulte-Römer explores in her chapter on female perceptions of dark skies. She brings together four experts to discuss gendered perspectives on night-time environments in both urban and rural settings. Their key insights challenge the prevailing assumption that women feel unsafe in outdoor spaces after dark and thus question the over-illumination of places. The discussion highlights that female fears, and human fears generally, are intensified by the lack of vision in the dark, media-shaped imaginations, and social dynamics.

Such examples challenge us to consider how we might better connect up divergent groups, practitioners, and policymakers in encompassing their concerns, ideas, and values in pursuit of a more inclusive approach to dark sky management and designation. This further summons up the notion that the global population can be construed as one vast dark skies community. Reconnecting ourselves with the sheer awe of dark skies, reminding ourselves of the fragile interdependencies between humans and other species, enabling ecosystems to flourish and minimising the waste of resources would rekindle our humanity. Being humble and wise enough to recognise these necessary transformations as progressive rather than regressive steps is crucial to their implementation. The level of cooperation, coordination, and collaboration needed may be daunting, but this does not make the situation any less urgent.

Schulte-Römer proposes the adoption of a cosmopolitics of dark skies that recognises the multiplicity of perspectives, including incommensurable values and incompatible views, through which we might account for tensions and complications in order to struggle towards solutions towards protecting our cosmic commons. By acknowledging the sheer historical and geographical diversity through which people have engaged with dark skies, reappraising the relationship between light and dark, extending and diversifying our encounters with darkness, and developing more inclusive approaches to understanding, managing, and designating dark sky places, we might intensify desires for their preservation as our awareness and knowledge accumulates. We conceive this book as adding to the momentum for substantive discussion and rapid action, as part of the campaign to move away from over-illumination and towards a renewed appreciation of the crepuscular worlds that emerge after nightfall.

References

Attlee, J. (2011) *Nocturne: A journey in search of moonlight*. London: Hamish Hamilton.

Barentine, J. C., Venkatesan, A., Heim, J., Lowenthal, J., Kocifaj, M. and Bará, S. (2023) 'Aggregate effects of proliferating low-Earth-orbit objects and implications for astronomical data lost in the noise', *Nature Astronomy*, 7, pp. 252–258. https://doi.org/10.1038/s41550-023-01904-2.

Beddington, E. (2023) 'Is light pollution making darkness a luxury?' *The Guardian*, 9 April. Available at: www.theguardian.com/lifeandstyle/2023/apr/09/goodbye-darkness-my-old-friend (Accessed: 19 June 2023).

Bennett, J. (2001) *The enchantment of modern life: Attachments, crossings and ethics*. Princeton, NJ: Princeton University Press.

Blair, A. (2016) *Sark in the dark: Wellbeing and community on the dark sky Island of Sark*. St Davids: Sophia Centre Press.

Bogard, P. (2013) *The end of night: Searching for natural darkness in an age of artificial light*. London: Fourth Estate.

Brigstocke, J. and Noorani, T. (2016) 'Posthuman attunements: Aesthetics, authority and the arts of creative listening', *GeoHumanities*, 2(1), pp. 1–7.

Brown, T. M., Brainard, G. C., Cajochen, C., Czeisler, C. A., Hanifin, J. P., Lockley, S. W., *et al.* (2022) 'Recommendations for daytime, evening, and nighttime indoor light exposure to best support physiology, sleep, and wakefulness in healthy adults', *PLoS Biology*, 20(3), p. e3001571. https://doi.org/10.1371/journal.pbio.3001571.

Concepto. (2012) *Lighting master plan of Rennes, France*. Available at: www.concepto.fr/portfolio_page/lighting-master-plan-rennes-france/.

Crary, J. (2013) *24/7: Terminal capitalism and the ends of sleep*. London: Verso.

Dark Source. (2020) Available at: www.dark-source.com (Accessed: 27 April 2023).

Davidson, P. (2015) *The last of the light: About twilight*. London: Reaktion.

Downey, C. (2021) 'Outside knowing: Accessing alterity in the nocturnal urban landscape', in Dortdivanlioglu, H. and Marratt, M. (eds.) *Proceedings of the ConCave: Divergence in architectural research*. Atlanta, GA: Georgia Institute of Technology, pp. 149–157.

Dunn, N. (2019) 'Dark futures: The loss of night in the contemporary city?', *Journal of Energy History / Revue d'Histoire de l'Énergie*, Special Issue: Light(s) and Darkness(es) / Lumière(s) et Obscurité(s) 1(2), pp. 1–27. http://energyhistory.eu/en/node/108.

Dunn, N. (2020) 'Dark design: A new framework for advocacy and creativity for the nocturnal commons', *International Journal of Design in Society*, 14(4), pp. 9–23.

Dunn, N. and Edensor, T. (eds.) (2020) *Rethinking darkness: Cultures, histories, practices*. London: Routledge.

Durand, M. (2021) 'The blue hour of the mystic North', in Chartier, D., Lund, K. and Johannesson, G. (eds.) *Darkness: The dynamics of darkness in the north*. Reykjavik: University of Iceland, pp. 209–221.

Edensor, T. (2014) 'The social life of the senses: Ordering and disordering the modern sensorium', in Howes, D. (ed.) *The cultural history of the senses: The modern age: 1920–2000*. Oxford: Berg, pp. 31–54.

Edensor, T. (2022) 'Dark designs: Creating shadow, gloomy spaces and enchanting light', in Sumartojo, S. (ed.) *Lighting design in shared public spaces*. London: Routledge, pp. 195–215.

Exmoor Dark Skies Festival. (2023) Available at: www.exmoor-nationalpark.gov.uk/enjoying/stargazing/dark-skies-festival (Accessed: 24 June 2023).

Ffrench-Constant, R. H., Somers-Yeates, R., Bennie, J., Economou, T., Hodgson, D., Spalding, A. and McGregor, P. K. (2016) 'Light pollution is associated with earlier tree budburst across the United Kingdom', *Proceedings of the Royal Society B*, 283. https://doi.org/10.1098/rspb.2016.0813.

Gandy, M. (2017) 'Negative luminescence', *Annals of the American Association of Geographers*, 107(5), pp. 1090–1107.

Gaston, K. J., Gardner, A. S. and Cox, D. T. C. (2023) 'Anthropogenic changes to the nighttime environment', *Bioscience*, 73(4), pp. 280–290. https://doi.org/10.1093/biosci/biad017.

Green Space, Dark Skies. (2022) Available at: https://greenspacedarkskies.uk (Accessed: 10 April 2023).

Hannah, M. (2013) 'Attention and the phenomenological politics of landscape', *Geografiska Annaler: Series B, Human Geography*, 95(3), pp. 235–250.

Jägerbrand, A. K. and Spoelstra, K. (2023) 'Effects of anthropogenic light on species and ecosystems', *Science*, 380(6650), pp. 1125–1130. https://doi.org/10.1126/science.adg3173.

Kocifaj, M., Kundracik, F., Barentine, J. C. and Bará, S. (2021) 'The proliferation of space objects is a rapidly increasing source of artificial night sky brightness', *Monthly Notices of the Royal Astronomical Society: Letters*, 504(1) (June), pp. L40–L44. https://doi.org/10.1093/mnrasl/slab030.

Koslofsky, C. (2011) *Evening's empire: A history of night in early modern Europe*. Cambridge: Cambridge University Press.

Lees, A. (2022) *After dark: Walking into the nights of Aotearoa*. Potton and Burton: Nelson.

Lowe, C. and Rafael, P. (2020) 'Designing with light and darkness', in Dunn, N. and Edensor, T. (eds.) *Rethinking darkness: Cultures, histories, practices*. London: Routledge, pp. 216–228.

Melbin, M. (1978) 'Night as frontier', *American Sociological Review*, 43(1), pp. 3–22.

Morgan-Taylor, M. (2023) 'Regulating light pollution: More than just the night sky', *Science*, 380(6650), pp. 1118–1120. https://doi.org/10.1126/science.adh7723.

Morris, N. (2011) 'Night walking: Darkness and sensory perception in a night-time landscape installation', *Cultural Geographies*, 18(3), pp. 315–342.

Narboni, R. (2017) 'Imagining the future of the city at night', *Architectural Lighting*. Available at: www.archlighting.com/projects/imagining-the-future-of-the-city-at-night_o.

Rancière, J. (2009) *The politics of aesthetics*. Translated by Rockhill, G. New York, NY: Continuum.

Ross, R., Burton-Page, M., van der Pol, J. and Férey, J. (2023) *LUCI declaration for the future of urban lighting*. Claix, France: Imprimerie du Pont de Claix.

Sánchez de Miguel, A., Bennie, J., Rosenfeld, E., Dzurjak, S. and Gaston, K. J. (2021) 'First estimation of global trends in nocturnal power emissions reveals acceleration of light pollution', *Remote Sensing*, 13, p. 3311. https://doi.org/10.3390/rs13163311.

Smith, K. T., Lopez, B., Vignieri, S. and Wible, B. (2023) 'Losing the darkness', *Science*, 380(6650), pp. 1116–1117. https://doi.org/10.1126/science.adi4552.

Soga, M. and Gaston, K. J. (2016) 'Extinction of experience: The loss of human-nature interactions', *Frontiers in Ecology and the Environment*, 14, pp. 94–101.

Spark, J. (2021) 'Beyond edges: Darkness and the fluxing field', in Chartier, D., Lund, K. and Johannesson, G. (eds.) *Darkness: The dynamics of darkness in the North*. Reykjavik: University of Iceland, pp. 23–41.

Wills, D. (2015) *At night: A journey round Britain from dusk to dawn*. Basingstoke: AA Publishing.

Yates, C. (2012) *Nightwalk: A journey to the heart of nature*. London: Collins.

Zielinska-Dabkowska, K. M. (2022) 'Healthier and environmentally responsible sustainable cities and communities. A new design framework and planning approach for urban illumination', *Sustainability*, 14, p. 14525. https://doi.org/10.3390/su142114525.

Index

Note: numbers in *italics* indicate a figure

Taylor & Francis Group
an **informa** business

Taylor & Francis eBooks

www.taylorfrancis.com

A single destination for eBooks from Taylor & Francis
with increased functionality and an improved user
experience to meet the needs of our customers.

90,000+ eBooks of award-winning academic content in
Humanities, Social Science, Science, Technology, Engineering,
and Medical written by a global network of editors and authors.

TAYLOR & FRANCIS EBOOKS OFFERS:

A streamlined
experience for
our library
customers

A single point
of discovery
for all of our
eBook content

Improved
search and
discovery of
content at both
book and
chapter level

REQUEST A FREE TRIAL
support@taylorfrancis.com

Routledge
Taylor & Francis Group

CRC Press
Taylor & Francis Group

For Product Safety Concerns and Information please contact our EU
representative GPSR@taylorandfrancis.com
Taylor & Francis Verlag GmbH, Kaufingerstraße 24, 80331 München, Germany

www.ingramcontent.com/pod-product-compliance
Lightning Source LLC
Chambersburg PA
CBHW052121230326
41598CB00080B/3927

9 781032 528038